HYDRODYNAMIC SCALES OF INTEGRABLE MANY-BODY SYSTEMS

HYDRODYNAMIC SCALES OF INTEGRABLE MANY-BODY SYSTEMS

Herbert Spohn
Technical University of Munich, Germany

World Scientific

NEW JERSEY · LONDON · SINGAPORE · BEIJING · SHANGHAI · HONG KONG · TAIPEI · CHENNAI · TOKYO

Published by

World Scientific Publishing Co. Pte. Ltd.

5 Toh Tuck Link, Singapore 596224

USA office: 27 Warren Street, Suite 401-402, Hackensack, NJ 07601

UK office: 57 Shelton Street, Covent Garden, London WC2H 9HE

Library of Congress Control Number: 2024930690

British Library Cataloguing-in-Publication Data
A catalogue record for this book is available from the British Library.

HYDRODYNAMIC SCALES OF INTEGRABLE MANY-BODY SYSTEMS

ISBN 978-981-12-8352-9 (hardcover)
ISBN 978-981-12-8353-6 (ebook for institutions)
ISBN 978-981-12-8354-3 (ebook for individuals)

For any available supplementary material, please visit
https://www.worldscientific.com/worldscibooks/10.1142/13600#t=suppl

Typeset by Stallion Press
Email: enquiries@stallionpress.com

To my grandson Lio Spohn for his continuing support.

Preface

In August 1974, for the first time, I attended a summer school, which happened to be number three in the series on "Fundamental Problems in Statistical Mechanics". The school took place at the Agricultural University of Wageningen. We were approximately 70 participants from 25 countries. The school lasted nearly three weeks with 11 lecture courses, each four hours long, and various more specialized seminars. As to be expected, present were the big shots, as Eddy Cohen who with moderate success tried to slow down the speaker by posing questions. Besides lectures, the truly exciting part of the school was meeting fellow youngsters who had similar interests and struggled more or less with the same difficulties.

At the time, critical phenomena and the just invented RG methods were the overwhelming topic. Fortunately, the Dutch physics community has a long tradition in Statistical Mechanics and therefore a wide range of topics were covered. I vividly recall the lectures by Nico van Kampen on "Stochastic differential equations". Joe Ford lectured on "The statistical mechanics of classical analytic dynamics", the early days of deterministic chaos. In fact, at the end of his lectures, Joe mentioned the Toda lattice describing the just confirmed integrability. Perhaps I should have listened with more care. The strongest impact had the lectures presented by Piet Kasteleyn on "Exactly solvable lattice models", explaining the fascinating link between equilibrium statistical mechanics and integrable models from quantum many-body physics.

My second encounter with integrable systems is related to the study of Dyson Brownian motion, which is an integrable stochastic particle system. At the time, to me, the model was an intriguing example of the hydrodynamics of a many-particle system with long-range forces. The third encounter was triggered by the KPZ revolution, which brought me in contact with further corners of integrable systems. Around 2016, I first learned about the activities investigating the hydrodynamic scales for integrable quantum many-body systems. I could not resist. Of course, major insights had been accomplished already. But, apparently, classical integrable many-particle systems were in a state of dormancy. This is how my enterprise got started.

Acknowledgments

During the ongoing project, I had many insightful comments and good advice. Gratefully acknowledged are Mark Adler, Amol Aggarwal, Vir Bulchandani, Xiangyu Cao, Kedar Damle, Avijit Das, Percy Deift, Jacopo De Nardis, Atharv Deokule, Abhishek Dhar, Maurizio Fagotti, Pablo Ferrari, Patrik Ferrari, Chiara Franceschini, Tamara Grava, Alice Guionnet, David Huse, Thomas Kappeler, Karol Kozlowski, Thomas Kriecherbauer, Manas Kulkarni, Anupam Kundu, Aritra Kundu, Gaultier Lambert, Joel Lebowitz, Guido Mazzuca, Ken McLaughlin, Christian Mendl, Pierre van Moerbecke, Joel Moore, Fumihiko Nakano, Neil O'Connell, Stefano Olla, Lorenzo Piroli, Balázs Pozsgay, Michael Prähofer, Sylvain Prolhac, Tomaz Prosen, Keiji Saito, Makiko Sasada, Tomohiro Sasamoto, Naoto Shiraishi, Jörg Teschner, Khan Duy Trinh, Simone Warzel, and Takato Yoshimura.

Special thanks are due to Benjamin Doyon. Our encounter at Pont-à-Mousson is well remembered.

When working on the manuscript, I had the opportunity to attend long programs at the Mathematical Sciences Research Institute in Berkeley, the Galileo Galilei Institute in Firenze, the Newton Institute in Cambridge, and the International Center for Theoretical Sciences in Bengaluru, in chronological order. This generous hospitality is highly appreciated.

Herbert Spohn
Munich, February 2024

Contents

Chapter 1

Overview

Hydrodynamics is based on the observation that the motion of a large assembly of strongly interacting particles is constrained by local conservation laws. As a result, local equilibrium is established over an initial time span to be followed by a much longer time window when local equilibrium parameters are governed by the hydrodynamic evolution equations. The initial time span could shrink to microscopic times when the system starts out already in local equilibrium. It is a matter of fact that a vast amount of interesting physics is covered by the hydrodynamic approach. Historically, the best known examples are simple fluids for which hydrodynamics is synonymous with fluid dynamics.

Already, for simple fluids, the hydrodynamic approach carries the seed for further extensions since the equilibrium phase diagram is richly structured. The most common is the occurrence of a liquid–gas phase transition. This discrete order parameter has now to be added as a further parameter characterizing local equilibrium. For example, gas and fluid phases may spatially coexist and the respective interface is then an additional slow degree of freedom, to be included in the macroscopic dynamics.

Close to critical points, conventional hydrodynamics has to be augmented by more refined theories. At lower temperatures, generically a solid phase stabilizes. Due to slow relaxation of solids, for the dynamics of the solid–gas interface, mostly non-hydrodynamic modeling is used. Bosonic particles at low temperatures will form a condensate. One then employs a hydrodynamic two-fluid model, which governs the superfluid interacting with the normal fluid. Going beyond short-range interactions, magneto-hydrodynamics describes the motion of fluids made up of charged particles, also including the Maxwell field as additional dynamical degrees of freedom.

Relativistic hydrodynamics becomes relevant for extreme events such as the formation of superdense neutron stars, relativistic jets, and Gamma ray bursts. Each topic mentioned is part of a vast enterprise with ongoing research.

Approximately seven years ago, a novel item was added to our list under the name of *generalized hydrodynamics* (GHD). Physically, perhaps not as far reaching as other areas mentioned, generalized hydrodynamics relies on an amazing twist. The novel topic is concerned with integrable many-particle models for which the number of conserved fields is proportional to system size, in sharp contrast to the models listed before which have only a few conserved fields, for example, number, momentum, energy, plus broken symmetries in case of simple fluids. At first glance, the mere idea of a hydrodynamic description of the time evolution of such an integrable system sounds like an intrinsic contradiction. After all, establishing local equilibrium relies on chaotic dynamics which is just the opposite of integrability. But the huge number of degrees of freedom helps. Since integrable many-particle systems have an extensive number of local conservation laws, local equilibrium must now be characterized by a correspondingly large number of chemical potentials. It is this feature which is called "generalized". In the limit of infinite system size, the hydrodynamic fields are labeled by a parameter taking integer values, $n = 0, 1, \ldots$, or possibly by more complicated labeling schemes. As a consequence, writing down the coupled set of hyperbolic conservation laws is already a major obstacle.

The notion of *generalized Gibbs ensemble* (GGE) was introduced already somewhat earlier and studied systematically in a related context, known as quantum quench. But the issue is generic. One starts from a spatially homogeneous random state and wants to identify the random state reached after a long time. More physically, one prepares a homogeneous state of a particular Hamiltonian dynamics and then abruptly changes the dynamics (the quench). If the quench dynamics is not integrable, generically, one expects the system to thermalize with parameters determined by the conserved fields when averaged over the initial state. But an integrable system has many conserved fields and the final state will depend on an extensive set of parameters. Such asymptotic states are called GGE.

For many-particle systems to be integrable requires fine-tuned interactions. Nevertheless, the list is not so short. The first and still much studied model is the Lieb–Liniger δ-Bose gas from 1963. The Toda lattice was discovered in 1967 and its integrability was firmly established seven years later through the construction of a Lax matrix. Further examples are the XXZ spin chain, the one-dimensional spin-$\frac{1}{2}$ Fermi–Hubbard model, the classical particle models of Calogero, and continuum wave equations as Korteweg–de Vries, nonlinear Schrödinger, and sinh-Gordon.

In fact, the central goal of our notes is to argue the following:

> On a hydrodynamic scale, all integrable many-particle systems are structurally alike.

Given the diversity of microscopic models, such a claim is surprisingly bold. On the other hand, as to be discussed, the route to tackle the hydrodynamic scale will depend on the specific model. The precise meaning of our claim will unfold. But to provide at least a very preliminary glimpse, in all models under study, the two-body scattering shift will be a crucial piece of the hydrodynamic description.

As familiar from fluids, Euler equations refer to the ballistic scale, which is characterized by space and time to be of same order of magnitude. Formally, entropy is locally conserved. Transport properties arise at longer diffusive time scales and are included in the hydrodynamic equations through the Navier–Stokes correction. For integrable systems, the same distinctions apply, at least in principle. This topic is briefly touched upon in the very last chapter of our notes. Otherwise, hydrodynamics is understood as ballistic Euler-type scaling.

From a broader perspective, in deriving the equations governing the motion on the hydrodynamic scale, one faces several difficulties:

(i) For a given system, the local conservation laws have to be listed in terms of which GGEs can then be constructed. In a somewhat vague sense, this list has to be complete since hydrodynamically all fields are expected to be coupled to each other.

(ii) The structure of the generalized free energy has to be understood, including its first-order derivatives which are linked to the GGE averaged conserved fields.

(iii) To complete the hydrodynamic equations, one has to know the GGE averaged currents as a functional of the GGE averaged conserved fields.

My exposition is *not* a review, even though much of the relevant literature is cited. To establish a guiding backbone, the classical Toda lattice is discussed in considerable detail. Particularly introduced are two distinct strategies: (1) a closed system with a linearly varying pressure and (2) the canonical transformation to scattering coordinates. The first method will also be applied to the Ablowitz–Ladik discretized nonlinear Schrödinger equation and the second one serves well for the Calogero fluid. The key quantum models accounted for will be the Lieb–Liniger δ-Bose gas and the quantized Toda lattice, for both models relying on the Bethe ansatz as strategy.

Let me refrain from further comments on the content of my notes and rather turn to some remarks on the history of the subject. The one-dimensional system of classical hard rods was studied around 1970. The system is integrable, since in a collision, momenta are merely exchanged. Conserved is any one-particle sum function depending only on the momenta. Due to the hard core the hydrodynamic fields are nonlinearly coupled, in sharp contrast to an ideal gas. Jerry Percus first derived the hydrodynamic equations. In the 1980s, Roland Dobrushin and collaborators analyzed in much greater detail the time evolution of hard rods. They well understood the hydrodynamic perspective, including the issue of Navier–Stokes corrections. But at the time, no tools were available for handling more intricate models. In retrospect, the true simplification of hard rods is a two-particle scattering shift which is independent of the incoming quasiparticle velocities.

Another early line of research concerns the Korteweg–de Vries equation which is an integrable nonlinear wave equation accessible through the inverse scattering transform. In the mid-1990s, Vladimir E. Zhakarov studied a low-density gas of solitons and derived the respective kinetic equation for the spacetime dependence of the soliton counting function. The extension to a dense soliton gas, to say the respective hydrodynamic equations, has been obtained by Gennady El in 2003.

Generalized hydrodynamics as a systematic research activity relies on a breakthrough advance in 2016 independently by the two groups: O.A. Castro-Alvaredo, B. Doyon, and T. Yoshimura *and* B. Bertini, M. Collura, J. De Nardis, and M. Fagotti. They discovered a general scheme of how to write down the average currents, thereby covering classical field theories and quantum many-body systems. Only with such an input, the equations of generalized hydrodynamics could be written with confidence, herewith opening the door to applications of physical interest. Such detailed studies strongly support our claim that on a hydrodynamic scale, all integrable many-particle models look alike.

In condensed matter physics, the notion quantum many-body is widely used. Many-particle system is more natural for models from classical mechanics. Both notions are employed interchangeably and comprise also classical and quantum field theories.

How the text is structured: Our material is arranged in 15 chapters with sections varying in number. Longer subsections might be separated by boldface headers. No quotations are provided in the main text. Instead, at the end of each chapter, one finds "Notes and references", which roughly means a bibliography with extended comments. In addition, there are Inserts separated from the main text by ♦♦ *Header*. ⋯ ♦♦. Typically, an Insert deals with a closely related topic, which however can be touched upon only superficially. Also some more technical derivations have been shifted to Inserts. The idea is that at first reading, an Insert can be skipped, except for conventions on notation.

Notes and references

Preface

The Proceedings of the Wageningen summer school have been edited by E.D.G. Cohen (1975). My work on Dyson Brownian motion is published in Spohn (1987). The KPZ revolution is covered in many articles from which only the overviews Corwin (2012), Quastel and Spohn (2015), Spohn (2017), and Takeuchi (2018) are quoted.

Overview

A useful account on the developments prior to generalized hydrodynamics can be found in a special volume on "Quantum Integrability in Out-of-Equilibrium Systems" edited by Calabrese *et al.* (2016). Its central theme are quantum quenches starting from a spatially homogeneous initial state.

The notion "scattering shift" is convenient but less widely used. It refers to the fact that, when two particles undergo a scattering motion, each trajectory is asymptotically of the form $v_j t + \phi_j^{\pm}$, $j = 1, 2$, as $t \to \pm\infty$, i.e., free motion linear in time and on top a constant displacement as first order correction, which is the scattering shift. In quantum mechanical two-body scattering, the wave function for the relative motion is asymptotically of the form $\exp\left[\mathrm{i}(-kx + \frac{1}{2}k^2 t + \theta(k))\right]$. Then, a narrow wave packet, centered at k_0 in momentum space, travels in physical space with velocity k_0 and is displaced by $\theta'(k_0)$. θ is the phase shift, while its derivative, θ', is the scattering shift.

Percus (1969) wrote down what he called a kinetic equation. In his prior work, in collaboration with Lebowitz and Sykes (1968), the exact spacetime two-point function of hard rods in thermal equilibrium was obtained. Boldrighini *et al.* (1983) prove, under fairly general assumptions on the initial probability measure, the validity of the Euler equation for a system of hard rods under ballistic scaling. They also established that the hydrodynamic equation has smooth solutions. Dobrushin (1989) is his vision on hydrodynamic limits. In the context of the Korteweg–de Vries equation, Zakharov (1971) studied a low-density gas of solitons with Poisson distributed centers and statistically independent soliton velocities. He argued for a Boltzmann-type kinetic equation. The hydrodynamic extension to a dense soliton fluid has been accomplished in El (2003) and El and Kamchatnov (2005), see also El *et al.* (2011) and Carbone *et al.* (2016). Soliton-based hydrodynamics is explained in Chapter 10 with more references to be added. The upswing of generalized hydrodynamics can be traced back to the two independent seminal contributions by Castro-Alvaredo *et al.* (2016) and Bertini *et al.* (2016), in which a general scheme for the computation of GGE averaged currents is presented, GGE being the by now standard acronym for "generalized Gibbs ensemble". Thus Euler type equations could be written down based on convincing theoretical reasoning.

Chapter 2

Dynamics of the Classical Toda Lattice

As well known at the time when Morikazu Toda started his studies on the lattice with exponential interactions, shallow water waves in long channels have peculiar dynamical properties. One observes solitary waves and soliton collisions. In the latter, two incoming solitons emerge with their original shape after an intricate dynamical process. The Korteweg–de Vries (KdV) equation, a one-dimensional nonlinear wave equation, provides an accurate theoretical description of these phenomena. In 1967, Toda investigated whether discrete wave equations also might have solitary-type dynamics. A point in case is the standard wave equation with the harmonic lattice as its spatial discretization, both of them linear equations. With ingenious insight, Toda discovered that a lattice with exponential interactions, later known as Toda lattice, exhibits the same dynamical features as the KdV equation. It took another 7 years until it had been firmly established that the N-particle Toda lattice is indeed integrable with $N+1$ conservation laws.

In dimensionless form, the Hamiltonian of the Toda lattice reads

$$H_{\mathrm{to}} = \sum_{j \in \mathbb{Z}} \left(\tfrac{1}{2} p_j^2 + \mathrm{e}^{-(q_{j+1}-q_j)} \right), \tag{2.1}$$

where $j \in \mathbb{Z}$ is the particle label and (q_j, p_j) are position and momentum of the jth particle. One could introduce a particle mass, m, coupling strength, g, and decay parameter, γ, as $(p_j^2/2m) + g\exp(-\gamma(q_{j+1} - q_j))$ with $m, g, \gamma > 0$. But through rescaling spacetime, the standard form (2.1) is recovered. The Toda chain has no free parameters. Following Newton, the equations of motion are

$$\frac{\mathrm{d}}{\mathrm{d}t}q_j = p_j, \qquad \frac{\mathrm{d}}{\mathrm{d}t}p_j = \mathrm{e}^{-(q_j - q_{j-1})} - \mathrm{e}^{-(q_{j+1} - q_j)}, \qquad j \in \mathbb{Z}. \tag{2.2}$$

Physically, this equation can be viewed in two different ways: (i) The displacements $\{q_j(t), j \in \mathbb{Z}\}$ are regarded as the lattice discretization of a continuum wave field $\{q(x,t), x \in \mathbb{R}\}$. We call this the lattice, or field theory, picture. (ii) The fluid picture is to literally view $\{q_j(t), j \in \mathbb{Z}\}$ as positions of particles moving on the real line. However, the particles do not interact pairwise as would be the case for a real fluid. The Hamiltonian is not invariant under relabeling of particles. Toda was mostly thinking of a lattice discretization. But the fluid picture is easier to visualize. Of course,

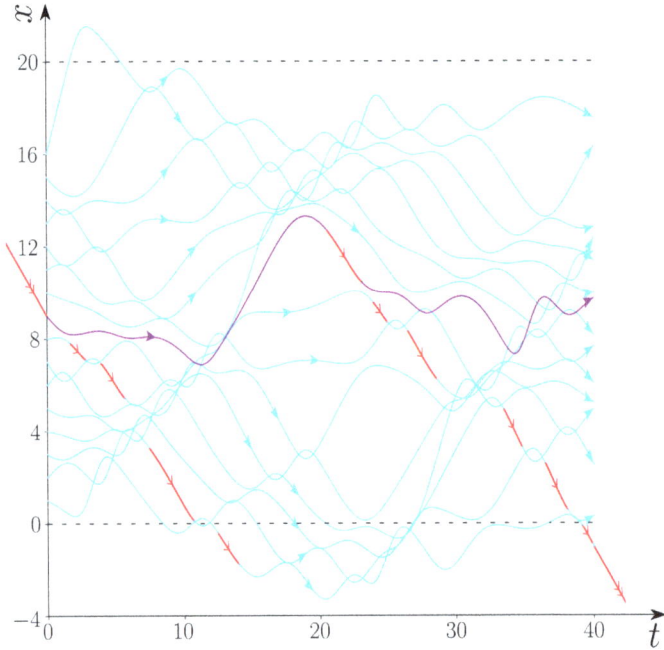

Fig. 2.1. Displayed are the trajectories, $q_j(t)$, of 16 Toda particles governed by $H_{\mathrm{cell},16}$ in a box of size $\ell = 20$. Very roughly, the stretch $\nu = 1.25$ and inverse temperature $\beta = 1$, which corresponds to the medium pressure $P = 0.7$. Clearly visible are two-particle collisions, but more complicated intertwined structures also develop. The trajectory of the ninth particle is shown in magenta. Quasiparticles are introduced by the condition to approximately maintain their velocity, except for collisions. One choice for a quasiparticle trajectory with the same starting point as the ninth particle is shown in red. From Spohn (2020a).

there is only a single set of equations of motion and one can switch back and forth between the two options.

At this stage, a review of the vast research on the Toda lattice can neither be supplied nor is it intended. We would have to refer to original research articles, monographs, reviews, and textbooks. However, it can be safely summarized that almost exclusively problems have been studied for which physically the chain is at zero temperature. Examples are multi-soliton solutions and the spatial spreading of local perturbations of an initially periodic particle configuration. In contrast, our focus are *random initial data* with an energy proportional to system size and away from a ground state energy. A paradigmatic set-up would be the thermal state at some non-zero temperature. For such an enterprise, novel techniques are required. In particular, the issue of large system size has to be properly understood.

As illustration in Figure 2.1 we show a numerical simulation of the particle dynamics and note that the widely used visualization as a sequence two-body collision seems to be of limited value.

2.1 Locally conserved fields and their currents

Our first task is to elucidate the integrable structure of the Toda lattice. For this purpose, we introduce the *stretch*

$$r_j = q_{j+1} - q_j, \tag{2.3}$$

also the free distance, or free volume, between particles j and $j + 1$, and the *Flaschka variables*

$$a_j = \mathrm{e}^{-r_j/2}, \quad b_j = p_j. \tag{2.4}$$

The stretch can have either sign, while $a_j > 0$. The *a*s and *b*s are conventional notation, but we will avoid the duplication of symbols by using only the momentum p_j. In terms of these variables, the equations of motion read

$$\frac{\mathrm{d}}{\mathrm{d}t}a_j = \tfrac{1}{2}a_j(p_j - p_{j+1}), \quad \frac{\mathrm{d}}{\mathrm{d}t}p_j = a_{j-1}^2 - a_j^2. \tag{2.5}$$

Hence, a_j couples to the right neighbor and p_j to the left one. In principle, we could have set $a_j = \tau \mathrm{e}^{-r_j/2}$ and $b_j = \tau p_j$, which amounts to a mere time change. Flaschka picked $\tau = \tfrac{1}{2}$. In the following, we will explain why $\tau = 1$ is singled out in our context.

For the purpose of thermodynamics, one first considers the lattice $[1, \ldots, N]$ with periodic boundary conditions. In Flaschka variables, they amount to the obvious condition

$$a_0 = a_N, \quad p_{N+1} = p_1. \tag{2.6}$$

This choice is also called the *closed* or *periodic* chain, for which the phase space is $\Gamma_N^\circ = (\mathbb{R}_+ \times \mathbb{R})^N$. Note that

$$\frac{\mathrm{d}}{\mathrm{d}t} \sum_{j=1}^{N} \log a_j^2 = 0, \tag{2.7}$$

since carrying out the time derivative yields a telescoping sum which vanishes by (2.6). Hence,

$$\sum_{j=1}^{N} r_j(t) = \ell, \tag{2.8}$$

with some constant ℓ, which can have either sign. Under this constraint, an equivalent description is to consider the infinite Toda chain and to impose the initial conditions

$$q_{j+N} = q_j + \ell, \quad p_{j+N} = p_j, \tag{2.9}$$

for some $\ell \in \mathbb{R}$ and all $j \in \mathbb{Z}$, which then holds at any time. Cutting the real line into cells, each of size ℓ, in every cell there are N particles and their dynamics is governed by the *cell Hamiltonian*:

$$H_{\mathrm{cell},N} = \sum_{j=1}^{N} \tfrac{1}{2} p_j^2 + \sum_{j=1}^{N-1} \mathrm{e}^{-(q_{j+1} - q_j)} + \mathrm{e}^{-(\ell - (q_N - q_1))}. \tag{2.10}$$

The positions are unconstrained. Total momentum is conserved and, under the dynamics generated by $H_{\mathrm{cell},N}$,

$$\sum_{j=1}^{N} q_j(t) = \sum_{j=1}^{N} q_j(0) + t \sum_{j=1}^{N} p_j(0). \tag{2.11}$$

The center of mass moves with constant velocity. The internal degrees of freedom are the stretches $r_1(t), \ldots, r_N(t)$ subject to the time-independent constraint (2.8). The stretches move in a potential which increases exponentially in all directions.

In the literature, periodic boundary conditions are often stated as $q_{N+1} = q_1$, which corresponds to the special case $\ell = 0$. Physically, the parameter ℓ is of crucial importance because through it the stretch per particle, ℓ/N, is controlled. In the fluid picture, the unit cell has length ℓ and contains N particles. The physical particle density is $N/|\ell|$. But often it is more natural to work with the signed particle density N/ℓ, which can be negative.

Out of the Flaschka variables, one forms the tridiagonal *Lax matrix*, L_N,

$$
L_N = \begin{pmatrix}
p_1 & a_1 & 0 & \cdots & & a_N \\
a_1 & p_2 & a_2 & \ddots & & 0 \\
0 & a_2 & p_3 & \ddots & & \vdots \\
\vdots & \ddots & \ddots & \ddots & & a_{N-1} \\
a_N & 0 & \cdots & & a_{N-1} & p_N
\end{pmatrix},
\tag{2.12}
$$

for $N \geq 3$, and its partner matrix

$$
B_N = \frac{1}{2} \begin{pmatrix}
0 & -a_1 & 0 & \cdots & & a_N \\
a_1 & 0 & -a_2 & \ddots & & 0 \\
0 & a_2 & 0 & \ddots & & \vdots \\
\vdots & \ddots & \ddots & \ddots & & -a_{N-1} \\
-a_N & 0 & \cdots & & a_{N-1} & 0
\end{pmatrix}.
\tag{2.13}
$$

L_N, B_N is called a *Lax pair*. While L is the common notation for a Lax matrix, the partner matrix is also denoted by A and M. Here, we follow the convention of Toda (1989). Clearly, L_N is symmetric and B_N skew symmetric, $(L_N)^{\mathrm{T}} = L_N$, $(B_N)^{\mathrm{T}} = -B_N$, with T denoting the transpose of a matrix. Later on for the adjoint of an operator also the more common * will be used. From the equations of motion (2.5), one verifies that

$$
\frac{\mathrm{d}}{\mathrm{d}t} L_N = [B_N, L_N]
\tag{2.14}
$$

with $[\cdot, \cdot]$ denoting the commutator, $[B_N, L_N] = B_N L_N - L_N B_N$. Since B_N is skew symmetric, $L_N(t)$ is isospectral to $L_N = L_N(0)$. Thus, the eigenvalues of L_N are conserved. Actually, any matrix B_N would do because

$$
\frac{\mathrm{d}}{\mathrm{d}t}\mathrm{tr}[(L_N)^n] = \sum_{j=0}^{n-1} \mathrm{tr}\big[(L_N)^j[B_N, L_N](L_N)^{n-1-j}\big] = \mathrm{tr}[B_N, (L_N)^n] = 0
$$

$$
\tag{2.15}
$$

by cyclicity of the trace for all n.

•• *Vector and matrix notation*: In our text, various N-vectors will appear. The standard notation $x \in \mathbb{R}^N$, $x = (x_1, \ldots, x_N)$, is adopted. The N-dimensional volume element is denoted by $\mathrm{d}^N x$. Also $x \in \mathbb{R}$ will be used. The distinction should be obvious from the context.

For matrices $\mathsf{A} = \{A_{i,j}\}_{i,j=1}^N$ with complex matrix elements, the transpose is denoted by $(\mathsf{A}^\mathsf{T})_{i,j} = A_{j,i}$ and the hermitian conjugate by $(\mathsf{A}^*)_{i,j} = \bar{A}_{j,i}$. Here, $\bar{z}$ is the complex conjugate of the complex number z.
••

•• *Notation for time-dependence*: For time-dependent quantities, as $X_j(t)$, we use the convention $X_j(0) = X_j$ and refer to X_j as time-zero field. While this notation is convenient, it might be ambiguous. For example, in (2.2), we should have used $q_j(t)$ with initial condition q_j. We anticipate that the exact meaning will be clear from the context.
••

•• *Phase spaces*: It is recommendable to keep track of phase spaces. We will use Γ as a generic symbol and Γ_n to indicate a phase space of dimension $2n$. More specifically, the notation is $\Gamma_N = \mathbb{R}^N \times \mathbb{R}^N$, $\Gamma_N^\circ = \mathbb{R}_+^N \times \mathbb{R}^N$, $\Gamma_N^\circ = \mathbb{R}_+^{(N-1)} \times \mathbb{R}^N$, and $\Gamma_N^\triangleright = \mathbb{W}_N \times \mathbb{R}^N$ with Weyl chamber $\mathbb{W}_N = \{q_1 < \cdots < q_N\}$.
••

Let us write the eigenvalue problem for L_N as

$$L_N \psi_\alpha = \lambda_\alpha \psi_\alpha \qquad (2.16)$$

with $\alpha = 1, \ldots, N$. Then, $\lambda_\alpha(a, p)$ is some function on phase space which does not change under the Toda time evolution. However, λ_α is a highly nonlocal function, in general. For example, considering the dependence of λ_α only on p_1 and $p_{N/2}$, this will not split into a sum as $g(p_1) + \tilde{g}(p_{N/2})$ even approximately. Physically more relevant are *local conservation laws*. For the Toda lattice, they are easily obtained through forming the trace as

$$Q^{[n],N} = \mathrm{tr}\big[(L_N)^n\big] = \sum_{j=1}^N (\lambda_j)^n = \sum_{j=1}^N ((L_N)^n)_{j,j} = \sum_{j=1}^N Q_j^{[n],N}. \qquad (2.17)$$

The second identity confirms that $Q^{[n],N}$ is conserved and the fourth identity that the density $Q_j^{[n],N}$ is local. Indeed, taking $n < N/2$ and expanding as

$$Q_j^{[n],N} = \sum_{j_1=1}^N \cdots \sum_{j_{n-1}=1}^N (L_N)_{j,j_1}(L_N)_{j_1,j_2} \ldots (L_N)_{j_{n-1},j}, \qquad (2.18)$$

the density $Q_j^{[n],N}$ depends only on the variables $\{a_{j-i}, p_{j-i}, \dots, a_{j+i}, p_{j+i},$ $i = 0, \dots, n-1\}$, modulo N.

The two-sided infinite volume limit of L_N is the tridiagonal Lax matrix L and correspondingly the partner matrix B, which are now operators acting on the Hilbert space $\ell_2(\mathbb{Z})$ of square-integrable two-sided sequences over the lattice $\mathbb{Z}$. The Lax pair still satisfies

$$\frac{\mathrm{d}}{\mathrm{d}t} L = [B, L]. \tag{2.19}$$

Of course, $\mathrm{tr}[L^n]$ makes no sense, literally. But the infinite volume density

$$Q_j^{[n]} = L_{j,j} \tag{2.20}$$

is well defined. $Q_j^{[n]}$ is a finite polynomial in the variables $\{a_i, p_i, |i - j| \le n - 1\}$, whose structure can be grasped more easily by using the random walk expansion derived from the $N = \infty$ version of (2.18). The walk is on $\mathbb{Z}$ with n steps of step size $0, \pm 1$, starting and ending at j. A step from i to i carries the variable p_i and the step either from i to $i + 1$ or from $i + 1$ to i carries the variable a_i. For a given walk, one forms the product along the path, which is a monomial of degree n, compare with Figure 2.2. $Q_j^{[n]}$ is then obtained by summing over all admissible walks. By translation invariance of the model, $Q_j^{[n]}$ and $Q_{j+i}^{[n]}$ are identical polynomials, except that the particle labels are shifted by i. Just as for the Hamiltonian (2.1), formally we still write

$$Q^{[n]} = \sum_{j \in \mathbb{Z}} Q_j^{[n]}, \tag{2.21}$$

where the superscript $[n]$ ranges over positive integers.

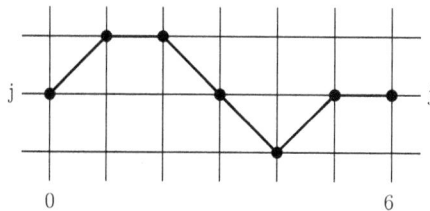

Fig. 2.2. A random walk path contributing to $Q_j^{[6]}$ and, according to the rules, carrying the weight $a_j p_{j+1} a_j a_{j-1} a_{j-1} p_j = a_{j-1}^2 a_j^2 p_j p_{j+1}$.

As remarked already, in addition, the stretch is locally conserved, which carries the label 0,

$$Q_j^{[0]} = r_j, \quad Q^{[0],N} = \sum_{j=1}^{N} r_j. \tag{2.22}$$

The three lowest-order fields have an immediate physical interpretation as stretch, momentum, and energy density,

$$Q_j^{[0]} = r_j, \quad Q_j^{[1]} = p_j, \quad \tfrac{1}{2} Q_j^{[2]} = \tfrac{1}{2}(p_j^2 + a_j^2 + a_{j-1}^2). \tag{2.23}$$

Obviously, there cannot be physical names for higher orders. In the community of quantum integrable systems, $Q^{[n]}$ is called the nth *conserved charge* or merely the nth *charge*, which serves as a concise notion but carries no specific physical meaning. We will use local conserved field and local charge interchangeably.

Since the Toda Hamiltonian has a local energy density, any locally conserved field must satisfy a continuity equation, in other words, the lattice version of a local conservation law. We consider the infinite lattice and compute

$$\frac{\mathrm{d}}{\mathrm{d}t} Q_j^{[n]} = (BL^n - L^n B)_{j,j} = a_{j-1}(L^n)_{j,j-1} - a_j(L^n)_{j+1,j} = J_j^{[n]} - J_{j+1}^{[n]}. \tag{2.24}$$

Hence, $J_{j+1}^{[n]}$ is the current of the nth conserved field from j to $j+1$ and $J_j^{[n]}$ the current from $j-1$ to j. Defining the lower triangular matrix $L^\downarrow$ by $(L^\downarrow)_{j+1,j} = a_j$ for all j and $(L^\downarrow)_{i,j} = 0$ otherwise, a more concise expression is

$$J_j^{[n]} = (L^n L^\downarrow)_{j,j}, \tag{2.25}$$

$n = 1, 2, \ldots$. For the stretch,

$$\frac{\mathrm{d}}{\mathrm{d}t} Q_j^{[0]} = -p_j + p_{j+1}, \quad J_j^{[0]} = -p_j, \tag{2.26}$$

and hence

$$J_j^{[0]} = -Q_j^{[1]}. \tag{2.27}$$

This innocent looking equation will have surprising consequences.

•• *Ambiguity of densities*: In the way just presented, the densities, both for field and current, seem to be unique. This is not the case, however.

We have made a particular choice which will be useful when investigating the hydrodynamic scale of the Toda lattice. A further common scheme is to require the density to depend on a minimal number of lattice sites. In terms of the random walk representation described above, this amounts to a density $\tilde{Q}_j^{[n],N}$ defined by summing over all closed admissible paths with minimum equal to j. As an example, for $n = 4$, the minimal version of the density is given by

$$\tilde{Q}_j^{[4]} = p_j^4 + 4(p_j^2 + p_j p_{j+1} + p_{j+1}^2)a_j^2 + 2a_j^4 + 4a_j^2 a_{j+1}^2. \tag{2.28}$$

Rather than trying to dwell on generalities, we illustrate the issue by considering the energy density, $n = 2$. From (2.18), we have

$$Q_j^{[2]} = p_j^2 + a_{j-1}^2 + a_j^2 \tag{2.29}$$

with the current density

$$J_j^{[2]} = a_{j-1}^2(p_{j-1} + p_j). \tag{2.30}$$

The minimal version of the energy density would be

$$\tilde{Q}_j^{[2]} = p_j^2 + 2a_j^2, \tag{2.31}$$

having the current density

$$\tilde{J}_j^{[2]} = 2a_{j-1}^2 p_j. \tag{2.32}$$

At infinite volume, we consider the spatial sums $\sum_{j=1}^{N} Q_j^{[2]} = Q^{[2],N\infty}$ and $\sum_{j=1}^{N} \tilde{Q}_j^{[2]} = \tilde{Q}^{[2],N\infty}$. They differ only by a boundary term and hence the spatial average $N^{-1}(Q^{[2],N\infty} - \tilde{Q}^{[2],N\infty}) = \mathcal{O}(1/N)$. On the other hand, the corresponding total currents, $J^{[2],N\infty}$ and $\tilde{J}^{[2],N\infty}$, differ by $\mathcal{O}(N)$, but

$$J_j^{[2]} - \tilde{J}_j^{[2]} = \frac{\mathrm{d}}{\mathrm{d}t}(a_{j-1})^2. \tag{2.33}$$

The difference is a total time derivative and thus vanishes when averaged over a time-stationary probability measure, e.g., thermal equilibrium. As a consequence, while there is ambiguity on the microscopic scale, upon averaging over large spacetime cells, this amounts to only small surface-type correction terms. In particular, the hydrodynamic equations for the Toda lattice do not depend on the particular choice of microscopic densities. ◆◆

2.2 Action–angle variables, notions of integrability

2.2.1 *Conventional integrability of the Toda lattice*

A cornerstone of Hamiltonian dynamics is the abstract characterization of integrable systems. Just to recall, given is some Hamiltonian, H, on a phase space Γ_n of dimension $2n$. The dynamics generated by H is called integrable, if there are n differentiable functions on phase space, the *action variables* $\{I_j, j = 1, \ldots, n\}$, which have the following properties: (i) They are conserved, which means that the Poisson brackets $\{I_j, H\} = 0$. (ii) They are in involution, i.e., $\{I_i, I_j\} = 0$ for all $i, j = 1, \ldots, n$. (iii) $I_1, \ldots, I_n$ span a n-dimensional hyper-surface in Γ_n, which is compact and connected. In particular, no scattering orbits are permitted. The hypersurface is assumed to be invariant under the Hamiltonian flow generated by I_j, for all j. So to speak, as regards the dynamics generated by I_j, the hypersurface has no boundary. Then, the Arnold–Liouville theorem states that there exists a canonical transformation to action variables $(I_1, \ldots, I_n)$ and the canonically conjugate *angle variables* $(\vartheta_1, \ldots, \vartheta_n) \in \mathbb{T}^n$, $\mathbb{T}^n$ the n-dimensional torus, such that the transformed Hamiltonian $\tilde{H}$ depends only on I. In these variables, the dynamics trivializes as

$$\frac{\mathrm{d}}{\mathrm{d}t}\vartheta_j = \omega_j, \qquad \omega_j = \partial_{I_j}\tilde{H}(I) \tag{2.34}$$

for $j = 1, \ldots, n$. The angle ϑ_j moves on the unit torus with frequency ω_j.

    This characterization applies also to the closed Toda chain. We consider a lattice of N sites, the phase space $\Gamma_N^\circ = (\mathbb{R}_+ \times \mathbb{R})^N$, and the evolution (2.5) in terms of the Flaschka variables $a = (a_1, \ldots, a_N)$ and momenta $p = (p_1, \ldots, p_N)$. These are not canonical variables. However, instead of the usual Poisson bracket, one can introduce a non-standard Poisson bracket by first defining the $N \times N$ matrix

$$A_N = \frac{1}{2}\begin{pmatrix} -a_1 & 0 & \cdots & 0 & a_N \\ a_1 & -a_2 & 0 & \ddots & 0 \\ 0 & a_2 & -a_3 & \ddots & \vdots \\ \vdots & \ddots & \ddots & \ddots & 0 \\ 0 & \cdots & 0 & a_{N-1} & -a_N \end{pmatrix} \tag{2.35}$$

and adjusting the Poisson bracket to

$$\{f, g\} = \langle \nabla_p f, A_N \nabla_a g \rangle - \langle \nabla_p g, A_N \nabla_a f \rangle \tag{2.36}$$

with $\langle \cdot, \cdot \rangle$ denoting the inner product in $\mathbb{R}^N$. For the usual Poisson bracket, A_N would be the identity matrix. In Flaschka variables,

$$H_{\text{to},N} = \sum_{j=1}^{N} \left(\tfrac{1}{2} p_j^2 + a_j^2 \right) \tag{2.37}$$

and the equations of motion (2.5) can be written in Hamiltonian form as

$$\frac{\mathrm{d}}{\mathrm{d}t} p_j = \{p_j, H_{\text{to},N}\}, \quad \frac{\mathrm{d}}{\mathrm{d}t} a_j = \{a_j, H_{\text{to},N}\}. \tag{2.38}$$

The matrix A_N is of rank $N-1$, the oblique projection for the eigenvalue 0 being

$$|a_1^{-1}, a_2^{-1}, \ldots, a_N^{-1}\rangle\langle 1, 1, \ldots, 1|, \tag{2.39}$$

i.e., the corresponding left eigenvector of A_N equals $-2\nabla_a Q^{[0],N}$ and the right one $\nabla_p Q^{[1],N}$. The Poisson structure (2.36) is degenerate. But it can be turned non-degenerate simply by fixing the two conservation laws $Q^{[0],N} = c_0$ and $Q^{[1],N} = c_1$ with an arbitrary choice of the real parameters c_0, c_1. The new phase space becomes Γ_{N-1}° and the dynamical evolution equations involve only the variables $(a_1, \ldots, a_{N-1}, p_1, \ldots, p_{N-1})$, which are a Hamiltonian system with a non-degenerate Poisson bracket structure. The phase space for the action–angle variables can be chosen as Γ_{N-1} with coordinates $(x_1, \ldots, x_{N-1}, y_1, \ldots, y_{N-1})$ and corresponding action–angle variables (ϑ, I), with $I_j \in \mathbb{R}_+, \vartheta_j \in [0, 2\pi]$, $j = 1, \ldots, N-1$, as

$$x_j = \sqrt{I_j} \cos \vartheta_j, \quad y_j = \sqrt{I_j} \sin \vartheta_j, \tag{2.40}$$

which are known as global Birkhoff coordinates. As proved in 2008 by A. Henrici and T. Kappeler, there is a canonical transformation Φ : $\Gamma_{N-1}^{\circ} \to \Gamma_{N-1}$ such that the transformed Hamiltonian, H_{aa}, depends only on the action variables, $H_{\text{aa}} = H_{\text{aa}}(I_1, \ldots, I_{N-1})$. In fact, as for us crucial property, H_{aa} is a strictly convex, real-analytic function. This means that the phases $\omega_j = \partial_{I_j} H_{\text{aa}}$ are incommensurate Lebesgue almost surely. In other words, H_{aa} has no linear pieces, as would be the case for a system of harmonic oscillators. The Toda lattice is phase-mixing: starting with some probability measure on Γ_{N-1}° with a continuous density function, in the long time limit, the density will become uniform on almost every torus of dimension $N-1$ with an amplitude computed from the initial density. The "almost every" is required because of tori with commensurate frequencies, which however form a set of Lebesgue measure zero.

♦♦ *Pitfalls of classical integrability*: We consider particles on the real line governed by the standard Hamiltonian

$$H_{\mathrm{mec},N} = \sum_{j=1}^{N} \tfrac{1}{2}p_j^2 + \sum_{i,j=1, i<j}^{N} V_{\mathrm{mec}}(q_i - q_j). \qquad (2.41)$$

The mechanical interaction potential is assumed to be even, $V_{\mathrm{mec}}(x) = V_{\mathrm{mec}}(-x)$, and repulsive, $V'_{\mathrm{mec}}(x) < 0$ for $x > 0$. Furthermore, the potential decays at infinity such that $|V_{\mathrm{mec}}(x)| < x^{-\gamma}$ for large x with some $\gamma > 1$ and diverges at the origin, thereby ensuring that particles do not cross. Hence, the phase space equals $\Gamma_N^{\triangleright}$. Under such assumptions, it is proved that asymptotic momenta exist,

$$\lim_{t\to\pm\infty} p_j(t) = p_j^{\pm}, \qquad (2.42)$$

and also the respective scattering shifts

$$\lim_{t\to\pm\infty} q_j(t) - p_j^{\pm}t = \phi_j^{\pm} \qquad (2.43)$$

as real-valued functions of the initial conditions. This defines the scattering map $\Phi^{-1} : (q,p) \mapsto (p^{\pm}, \phi^{\pm})$. The asymptotic momenta are ordered as $p_1^+ < \cdots < p_N^+$ and the scattering map is one-to-one on $\Gamma_N^{\triangleright}$. The scattering map is canonical, which implies that Poisson brackets are conserved. In particular,

$$\{p_i^+, p_j^+\} = 0 \qquad (2.44)$$

and

$$H_{\mathrm{mec},N} \circ \Phi = \sum_{j=1}^{N} \tfrac{1}{2}(p_j^+)^2. \qquad (2.45)$$

For the past, asymptotic momenta are anti-ordered and the corresponding properties hold as well.

Clearly, the family $\{p_j^+, j = 1, \ldots, N\}$ is in involution and conserved. As premature reaction, the mechanical system could be classified as integrable. However, the Arnold–Liouville theorem does not apply. Instead of quasi-periodic motion on tori, according to (2.45), in action–angle variables, the dynamics trivializes as $\phi_j^+ + p_j^+ t$. For periodic boundary conditions, integrability can be tested through molecular dynamics simulations. For a generic choice of V_{mec}, the particle system will thermalize in the long time limit. Such dynamics is very different from the one for the Toda fluid on a ring. N-body integrability requires fine-tuning of the interaction potential. One option would be to ask for a Lax matrix. Then, under our conditions, it is known that the only choices are $V_{\mathrm{mec}}(x) = \sinh^{-2}(x)$ and $V_{\mathrm{mec}}(x) = x^{-2}$, see Chapter 11 for more details. But this option is *ad hoc* without link to standard definitions. In the following section, we will argue that for hydrodynamic purposes, the natural defining property is a quasilocal density for the conserved fields. ◆◆

2.2.2 *Hydrodynamic perspective on integrability*

The Toda lattice is a very peculiar dynamical system in the sense that it is integrable for every system size N, which we call *integrable many-particle* or *integrable many-body*. Now, the large N limit is in focus and from a physics perspective, the conventional definition of integrability might have to be reconsidered. This is even more urgent, since the naive extension of classical integrability to quantum systems fails. A further desideratum would be a notion referring directly to the infinite lattice. For hydrodynamics, the central building blocks are local conservation laws, to be more precise, the Hamiltonian and the conservation laws are constructed from a strictly local density. We thus propose to call an infinitely extended system *nonintegrable*, if it admits only a few strictly local conservation laws. The system is called *integrable*, if it possesses an infinite number of linearly independent local conservation laws.

Starting from a strictly local density supported on an interval of m sites, the property to be the density of a conservation law refers to a phase space of dimension $2(n+\ell)$, in case the Hamiltonian density is supported on ℓ sites. The condition of being in involution is no longer mentioned, but it seems to hold in concrete examples, possibly after first adjusting either the classical phase space or the Poisson bracket. Note that local conservation laws have a linear structure, in the sense that the sum of two local conservation laws is again a local conservation law.

For the Toda lattice, the Lax matrix is tridiagonal implying strictly local conservation laws. A further integrable model, called Calogero fluid, are particles interacting through a $1/\sinh^2$ pair potential. This potential decays exponentially and the Lax matrix is fully occupied. Hence, the conserved fields of the Calogero fluid cannot be strictly local. Physically, the hydrodynamic scale has to include *quasilocal* fields with densities having exponential tails. However, depending on the model, the precise borderline could be a subtle issue.

⬩⬩ *Toda local conservation laws*: The Toda chain is integrable in the hydrodynamic sense with densities of the locally conserved fields stated in (2.20) and (2.22). However, as a stronger property, one would like to establish that there are no further local conservation laws. Currently, this is a conjecture and more studies would be needed. Still, a precise formulation is worthwhile. As before, periodic boundary conditions are understood. We assume some general density function f of support κ, in other words, $f(a_1, p_1, \ldots, a_\kappa, p_\kappa)$. Then, the shifted densities are $f_j(a, p) = f(a_{j+1}, p_{j+1}, \ldots, a_{j+\kappa}, p_{j+\kappa})$, $f_0 = f$, and the conditions for being a local conservation law read

$$Q^{[f],N} = \sum_{j=1}^{N} f_j, \quad \{Q^{[f],N}, H_{\text{to},N}\} = \sum_{j=1}^{N} \{f_j, H_{\text{to},N}\} = 0 \qquad (2.46)$$

for $N > 2\kappa$. If so, each of the Poisson brackets is a local function. By translation invariance the sum has to be telescoping and thus necessarily there exists a current function, J_j, of support of size $\kappa + 2$, such that $\{f_j, H_{\text{to},N}\} = J_j - J_{j+1}$. According to the already proven conventional integrability, there must be some function, G, such that

$$Q^{[f],N} = G(Q^{[0],N}, \ldots, Q^{[N],N}). \qquad (2.47)$$

Our conjecture claims that G is necessarily linear.

Conjecture: For fixed κ and N sufficiently large, there exist coefficients $c_0, c_1, \ldots, c_\kappa$ such that

$$Q^{[f],N} = \sum_{m=0}^{\kappa} c_m Q^{[m],N}. \qquad (2.48)$$

One argument in favor of the conjecture comes from a simple observation. Consider some locally conserved field, $Q^{[n],N}$. Then, $(Q^{[n],N})^2$ is also conserved but no longer local. The condition of locality should be strong enough to force a linear function in (2.47). ⬩⬩

The reader might find the evidence for linking integrability and conservation laws not particularly convincing. Agreed, but it should be noted that our definition translates one-to-one to quantum spin chains and also continuum quantum models. In the former case, there are two concrete results strongly supporting the hydrodynamic notion of integrability.

Considered is the XYZ spin chain with couplings J_x, J_y, J_z and external magnetic field, h, pointing in the z-direction. In terms of the Pauli spin-$\frac{1}{2}$ matrices, $\sigma^x, \sigma^y, \sigma^z$, the Hamiltonian reads

$$H_{\mathrm{XYZ}} = \sum_{j \in \mathbb{Z}} \left(J_x \sigma_j^x \sigma_{j+1}^x + J_y \sigma_j^y \sigma_{j+1}^y + J_z \sigma_j^z \sigma_{j+1}^z - h \sigma_j^z \right). \tag{2.49}$$

This model is integrable for $h = 0$ and for $h \neq 0$ in case $J_x = J_y$, the XXZ model with external field. The model is expected to be nonintegrable for any other choice of parameters. Now, for the case of non-zero coupling constants and $h \neq 0$, $J_x \neq J_y$, it is proved that there is only a single local conservation law, namely the Hamiltonian H_{XYZ} itself. This is, so to speak, the fully chaotic case. Only energy is transported and the chain thermalizes, in the sense that expectations of local observables converge to the thermal average in the long time limit.

On the other hand, for $h = 0$ and non-vanishing couplings, the model is integrable. Most efficiently, the strictly local conserved charges are computed through the boost operator. The second result states that, if the coupling constants are non-vanishing, then any local conservation law is a finite linear combination of the already known conservation laws. This is the precise analog of our conjecture. However, around 2014, for the XXZ chain with magnetic field, it was discovered that our notion of locality is indeed too restrictive. The XXZ chain possesses in addition quasilocal charges which have to be included in a hydrodynamic description, see Notes for details.

In passing, we note that the XXX chain at $h = 0$ is integrable and all three spin components, $\sigma_j^x, \sigma_j^y, \sigma_j^z$, are conserved. However, they do not commute with each other. There is still a tower of local charges, $Q_2, Q_3, \ldots$ in the usual notation, which commute with each other and with each spin-component. Strictly speaking, the involution property is violated.

Our discussion raises some difficult issues. From the available evidence, there is a dichotomy, either a few conservation laws or infinitely many. One does not know whether there is a deep reason behind or merely reflects the limited class of models studied. Personally, I believe in the first option. In this context, particularly intriguing are nearly integrable systems. For finite N, the KAM theorem provides information on the stability of the invariant tori, confirming the coexistence of chaotic and integrable regions in phase space. But in our definition, the limit $N \to \infty$ is taken first and integrable regions might be rare.

2.3 Scattering theory

In hydrodynamics, the system is confined and thus interactions persist without interruption, which in the long time limit then leads to some sort of statistical equilibrium. A dynamically distinct set-up is scattering: in the distant past, particles are in the incoming configuration, for which they are far apart and do not interact. A time span of multiple collision processes follows. In the far future, the outgoing particles move freely again. If the interaction potential is repulsive, no bound states can be formed. To illustrate the special features of scattering for integrable many-body systems, we first discuss a fluid consisting of hard rods, which will serve as an instructive example also later on.

2.3.1 *Hard rod fluid*

We consider N hard rods, rod length $a > 0$, moving on the real line. The Hamiltonian reads

$$H_{\mathrm{hr},N}^{\diamond} = \sum_{j=1}^{N} \tfrac{1}{2} p_j^2 + \sum_{j=1}^{N-1} V_{\mathrm{hr}}(q_{j+1} - q_j) \qquad (2.50)$$

with the hard rod potential $V_{\mathrm{hr}}(x) = \infty$ for $|x| < a/2$ and $V_{\mathrm{hr}}(x) = 0$ for $|x| \geq a/2$. Hard rods collide elastically with their two neighbors, except for the border particles with only one neighbor. Obviously, the system is integrable with one-particle sum functions, $\sum_{j=1}^{N} f(p_j)$, being conserved.

As written, hard rods interact with their nearest neighbors, just as for the Toda lattice. However, one could formally extend to all pairs (i, j) since interactions beyond nearest neighbor vanish because of the hard core. Fluid and lattice picture is discussed in detail in Chapter 5.

Since particles cannot cross, we order $q_1(t) < q_2(t) < \cdots < q_N(t)$. For sufficiently long times toward future and past,

$$q_j(t) = p_j^+ t + \phi_j^+, \quad t \to \infty, \qquad q_j(t) = p_j^- t + \phi_j^-, \quad t \to -\infty, \qquad (2.51)$$

where ϕ_j^+ is the forward in time and ϕ_j^- the backward *scattering shift*. When comparing with the point dynamics, $a \to 0$, one concludes

$$p_j^+ = p_{N-j+1}^-, \qquad p_N^- < \cdots < p_1^-, \qquad p_1^+ < \cdots < p_N^+. \qquad (2.52)$$

The *relative scattering shift* of particle j, κ_j, is defined by the deviation relative to the point particle dynamics. Hence,

$$\kappa_j = \phi_{N-j+1}^+ - \phi_j^-. \qquad (2.53)$$

Introducing the hard rod scattering shift

$$\phi_{\mathrm{hr}}(w) = -a, \tag{2.54}$$

and considering the intersection points arising from N straight spacetime lines, one concludes

$$\kappa_j = \sum_{i=1, i \neq j}^{N} \mathrm{sgn}(p_j^- - p_i^-) \phi_{\mathrm{hr}}(p_j^- - p_i^-) \tag{2.55}$$

when written for incoming momenta. A similar expression holds for outgoing momenta. As a hallmark of integrable many-body systems, the relative scattering shift is the properly signed sum of two-particle scattering shifts.

More intuitive is the notion of a *quasiparticle*, which maintains its velocity through a collision. Since for hard rods the collision time vanishes, a quasiparticle moves along a straight line interrupted by jumps of size *a* either to the right or to the left. Quasiparticles are ordered increasingly according to the ingoing particle configuration. Then, the relative scattering shift κ_j is the accumulated spatial shift of the jth quasiparticle.

Our definitions involve sign conventions which vary from author to author. As illustrated in Figure 2.3, by our rules, a negative scattering shift means that through a collision the two incoming particles get pushed apart relative to the free particle motion, while for a positive scattering shift, they get pulled closer. Hence, the two-particle scattering shift of the hard rod fluid equals $-a$, which corresponds to the relative scattering shift of particle 1 in case incoming momenta are ordered as $p_2 < p_1$.

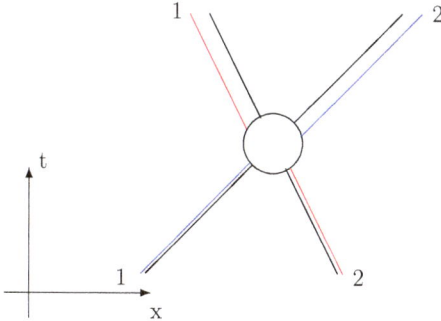

Fig. 2.3. Two-particle scattering. The numbers refer to particle labels. The red trajectory is quasiparticle 1 and blue quasiparticle 2. Quasiparticle 1 is shifted by ϕ_{12}, negative in the example, and quasiparticle 2 by $-\phi_{12}$.

For physical rods, the length is positive. But our rules make perfectly sense also for negative a. The two rods pass through each other until they reach the now negative distance a. At that moment, the velocities are exchanged. Since all rods have the same length, no ambiguities arise. Without much ado, the assumption $a > 0$ will be dropped. However, in our verbal explanations, we have to be more cautious.

2.3.2 Two-particle Toda lattice

Scattering is physically more intuitive in the fluid picture, i.e., particles move on the real line, also called the *open* chain. For two particles, the equations of motion are

$$\ddot{q}_1(t) = -e^{q_1 - q_2}, \quad \ddot{q}_2(t) = e^{q_1 - q_2}. \tag{2.56}$$

The relative motion, $q_2(t) - q_1(t)$, corresponds to a single particle subject to the potential $2e^{-x}$. We impose the asymptotic conditions $p_1(-\infty) = p_1^-$ and $p_2(-\infty) = p_2^-$ with $p_2^- < p_1^-$. Adjusting the initial time such that $q_2(0) - q_1(0)$ is at the turning point, i.e., $\dot{q}_2(0) - \dot{q}_1(0) = 0$, the solution to (2.56) becomes

$$q_1(t) = \tfrac{1}{2}(p_1^- + p_2^-)t - \log\left(\gamma^{-1}\cosh(\gamma t)\right),$$
$$q_2(t) = \tfrac{1}{2}(p_1^- + p_2^-)t + \log\left(\gamma^{-1}\cosh(\gamma t)\right), \tag{2.57}$$

with $\gamma = \tfrac{1}{2}(p_1^- - p_2^-) > 0$. The large time asymptotics is given by

$$q_1(t) = \begin{cases} p_1^- t + \log|p_1^- - p_2^-|, \\ p_2^- t + \log|p_1^- - p_2^-|, \end{cases} \quad q_2(t) = \begin{cases} p_2^- t - \log|p_1^- - p_2^-|, & t \to -\infty, \\ p_1^- t - \log|p_1^- - p_2^-|, & t \to \infty. \end{cases}$$
$$\tag{2.58}$$

Since quasiparticle one is shifted by $2\log|p_1^- - p_2^-|$ and quasiparticle two by $-2\log|p_1^- - p_2^-|$, we conclude that the Toda two-particle relative scattering shift is given by

$$2\log|p_1^- - p_2^-| = \phi_{\mathrm{to}}(p_1^- - p_2^-). \tag{2.59}$$

The scattering shift has no definite sign. For $|p_1^- - p_2^-| = 1$, the scattering shift vanishes. For $|p_1^- - p_2^-| < 1$, the scattering shift is negative, just as for hard rods. The trajectories of the two Toda particles look similar to the ones of hard rods, but the hard rod zero collision time is smeared to an exponential with rate γ. For $|p_1^- - p_2^-| > 1$, the scattering shift is positive. The trajectories spatially cross each other, still approaching their asymptotic motion exponentially fast.

2.3.3 N-particle Toda lattice

For N particles, the Hamiltonian of the open chain reads

$$H_{\text{to},N}^{\diamond} = \sum_{j=1}^{N} \tfrac{1}{2}p_j^2 + \sum_{j=1}^{N-1} e^{-(q_{j+1}-q_j)}, \tag{2.60}$$

where the superscript $\diamond$ is used to indicate the open chain. In Flaschka variables, the equations of motion become

$$\frac{\mathrm{d}}{\mathrm{d}t}a_j = \tfrac{1}{2}a_j(p_j - p_{j+1}), \tag{2.61}$$

$j = 1, \ldots, N-1$, and

$$\frac{\mathrm{d}}{\mathrm{d}t}p_j = a_{j-1}^2 - a_j^2, \tag{2.62}$$

$j = 1, \ldots, N$, with the boundary conditions $a_0 = 0$, $a_N = 0$. The Lax matrix $L_N^{\diamond}$ equals L_N except for $a_N = 0$, correspondingly for the partner matrix $B_N^{\diamond}$. The time evolution is encoded as

$$\frac{\mathrm{d}}{\mathrm{d}t}L_N^{\diamond} = [B_N^{\diamond}, L_N^{\diamond}]. \tag{2.63}$$

In general, modifying boundary conditions is likely to break integrability. But the open Toda chain is still integrable.

The time zero phase point is denoted by $(q, p) \in \Gamma_N$. Since particles repel, for sufficiently large t, one has $q_1(t) < \cdots < q_N(t)$ and $p_1(t) < \cdots < p_N(t)$. We order the eigenvalues of $L_N^{\diamond}$ as $\lambda_1 < \cdots < \lambda_N$, which defines the *Weyl chamber* $\mathbb{W}_N$. Then, in the limit $t \to \infty$,

$$q_j(t) \simeq \lambda_{N-j+1}t + \phi_j^+ \tag{2.64}$$

with the forward scattering shift ϕ_j^+. Correspondingly, in the past,

$$q_j(t) \simeq \lambda_j t + \phi_j^- \tag{2.65}$$

for $t \to -\infty$. The relative scattering shift is given by

$$\kappa_j = \sum_{i=1, i \neq j}^{N} \text{sgn}(\lambda_j - \lambda_i)\phi_{\text{to}}(\lambda_j - \lambda_i), \tag{2.66}$$

compare with (2.55) for which the same conventions are used. In view of Figure 2.1, this result is very surprising. Despite the intricate pattern of

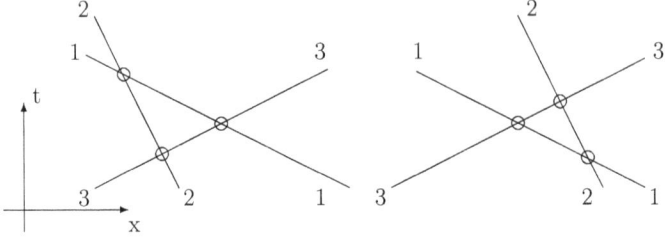

Fig. 2.4. Scattering shifts κ_j. The numbers refer to quasiparticle labels. In the left part the sequence of collisions is $\phi_{23}, \phi_{13}, \phi_{12}$ and in the right part $\phi_{12}, \phi_{13}, \phi_{23}$. Hence $\kappa_1 = \phi_{12} + \phi_{13}$, $\kappa_2 = -\phi_{12} + \phi_{23}$, and $\kappa_3 = -\phi_{23} - \phi_{13}$, independently of the particular sequence.

multiple collisions, at very long times, particles manage to have a scattering shift which is the weighted sum of two-particle scattering shifts. As one consequence of (2.66), the scattering shift does not depend on the order of collisions. For quantum mechanical many-body systems, this property is known as Yang–Baxter relation, see Chapter 14, and most commonly illustrated for three particles as in Figure 2.4.

We return to the scattering map Φ^{-1} defined through (2.64) and for notational simplicity denote ϕ^+ by ϕ, just dropping the superscript. As limit of canonical transformations, the scattering map $\Phi^{-1} : (q, p) \mapsto (\lambda, \phi)$ is symplectic, which means that the asymptotic momenta and the scattering shifts are canonical coordinates. In other words, as functions on Γ_N, the Poisson brackets read

$$\{\lambda_i, \lambda_j\} = 0, \quad \{\phi_i, \phi_j\} = 0, \quad \{\phi_i, \lambda_j\} = \delta_{ij}. \tag{2.67}$$

The variables (λ, ϕ) are the scattering analog of action–angle variables. Only angles vary over $\mathbb{R}$, rather than taking values on a torus. To make this distinction also verbally, one should speak of variables for asymptotic momenta and scattering shifts. But such practice becomes unwieldy and the common usage is *scattering coordinates* and more specifically *action–angle variables*, keeping in mind that a scattering situation is discussed.

Somewhat unexpectedly, for the Toda lattice, the scattering map can be made explicit. The formulas simplify by considering the inverse transformation. We define the map $\Phi : \Gamma_N^\triangleright \to \Gamma_N$, i.e., $\Phi : (\lambda, \phi) \mapsto (q, p)$ and $\Gamma_N^\triangleright = \mathbb{W}_N \times \mathbb{R}^N$. The map Φ is one-to-one, holomorphic, and given by

$$q_j = \log\left(\sigma_{N+1-j}/\sigma_{N-j}\right), \quad p_j = \left(\dot{\sigma}_{N+1-j}/\sigma_{N+1-j}\right) - \left(\dot{\sigma}_{N-j}/\sigma_{N-j}\right). \tag{2.68}$$

Here, $\sigma_0 = 1$ and

$$\sigma_k = \sum_{|I|=k} \exp\left(\sum_{i \in I} \phi_i\right) \prod_{i \in I, j \notin I} |\lambda_i - \lambda_j|^{-1} \qquad (2.69)$$

for $k = 1, \ldots, N$, where the first sum is over all subsets $I \subset \{1, \ldots, N\}$ of cardinality $|I| = k$. The symbol $\dot{\sigma}_k$ refers to the Poisson bracket with the transformed Hamiltonian,

$$\dot{\sigma}_k = \left\{\sigma_k, \sum_{j=1}^{N} \tfrac{1}{2}\lambda_j^2\right\}. \qquad (2.70)$$

Later on, we will use these expressions to compute the generalized free energy, see Section 9.2.

Notes and references

Section 2.0

The ground breaking discoveries are Toda (1967a,b). The second edition of the book is *Theory of Nonlinear Lattices*. Toda (1989) is still the most complete account up to 1989. Faddeev and Takhtajan (2007) is a widely used standard monograph on classical integrable systems. A somewhat more elementary introduction is Arutyunov (2019). Specifically, the Toda lattice is reviewed by Krüger and Teschl (2009). Closer to hydrodynamics is the study of a particular shock problem by Venakides *et al.* (1991). The 50 years anniversary volume edited by Bazhanov *et al.* (2018) provides a glimpse on research in vastly diverse directions.

Section 2.1

Based on explicit soliton solutions, Toda conjectured integrability. For the case of three particles, Ford *et al.* (1973) obtained very supporting numerical Poincaré plots. The integrals of motion in full generality were obtained by Hénon (1974) by a tricky enumerative argument. Hénon was worried about locality. Flaschka (1974) had the advantage of working at the Courant Institute, at which Peter Lax (1968) introduced his matrix in the context of the Korteweg-de Vries equation, see Section 10.1. Once the Lax matrix had been discovered, locally conserved fields are easily constructed. Independently, the Lax matrix for the Toda lattice has been reported by Manakov (1974). Apparently, at the time and later on, currents were hardly in focus, one exception being Shastry and Young (2010), who study the energy Drude weight in thermal equilibrium.

Section 2.2

To find out the canonically conjugate angles is a much more technical enterprise, which was accomplished in a series of papers by Henrici and Kappeler (2008a,b,c), starting from an early proposal by Flaschka and McLaughlin (1976), see Ferguson *et al.* (1982) for complimentary aspects. The pitfalls are also discussed in Ruijsenaars (1999), Section 5. Lucid proofs of the stated properties are presented by Hubacher (1989). As a further result, the additivity (2.66) is deduced when assuming the property $p_j^+ = p_{N+1-j}^-$. In spirit, this result ensures the existence of quasiparticles. However, the abstract proof does not tell us for which interaction potentials the assumption holds.

For quantum systems, the notion of integrability is controversial, since the naive transcription of the classical notion would mean that every eigenprojection of the Hamiltonian is conserved, in itself not such a helpful observation. We refer to Caux and Mossel (2011) for an exhaustive discussion. The link between integrability and local conservation laws has been mostly pushed by the quantum community, see Grabowski and Mathieu (1994, 1995) for early work. For classical systems, in general, this avenue still needs to be further developed. The mentioned results for the XYZ chain are prototypical for what one would like to achieve. The nonintegrable case is a result of Shiraishi (2019). The integrable case, $h = 0$, has been studied already by Grabowski and Mathieu (1994, 1995) with recent progress by Nozawa and Fukai (2020). Considering the XXZ chain with parameters $J_x = J_y = 1$, $J_z = \Delta$ and using only strictly local charges of the spin chain, one computes the spin Drude weight at zero magnetization, i.e., the persistent spin current, by using the Mazur formula. By spin inversion symmetry, this weight turns out to be identically 0. On the other hand, for $0 \leq \Delta < 1$, numerical evidence and exact steady state results for the boundary driven chain indicate that the Drude weight does not vanish. The puzzle is resolved by the construction of quasilocal conserved charges in Mierzejewski *et al.* (2015), see also related work Ilievski *et al.* (2015, 2016), for a broader perspective Doyon (2017), and the recent review Ilievski (2022). The Drude weight turns out to be nowhere continuous in its dependence on Δ.

The relation between integrability and conservation laws has been investigated also in the context of $1 + 1$-dimensional quantum field theories. Claimed is indeed a dichotomy in the following sense: If beyond the conservation of energy and momentum there is a single higher-order locally conserved charge, then necessarily the theory is integrable, in the sense of possessing infinitely many conservation laws, see Coleman and Mandula (1967), Iagolnitzer (1978a,b), Parke (1980), and Doyon (2008).

Section 2.3

In a beautiful piece of analysis, Moser (1975) proves the scattering shift for the N-particle Toda lattice. An account of his work can be found in Toda (1989). From a different perspective, a more recent discussion are the notes of Deift *et al.* (2019) based on his course at the Courant Institute in Spring 2019. The action–angle map Φ was established by Ruijsenaars (1990), where also the stated properties are proved. Asymptotic momenta equal eigenvalues of the Lax matrix. In spirit, there is also a corresponding algebraic identity for the scattering shifts, which has been constructed in analogy to more accessible models. But to check directly the validity of the Poisson bracket relations in (2.67) seems to be completely out of reach. Thus, as a major difficulty, one first has to establish agreement between algebraic approach and scattering map. A more recent point of view is developed by Fehér (2013).

The distinction between integrable and nonintegrable scattering has been studied in detail for one-particle systems. An example is the four hill potential $V_{\mathrm{fh}}(q_1, q_2) = q_1^2 q_2^2 \exp\left(- q_1^2 - q_2^2\right)$. Chaotic scattering is reviewed by Seoane and Sanjuán (2013).

Chapter 3

Static Properties

Considering the Toda lattice with periodic boundary conditions and highly excited initial conditions q, p, one would expect that in the long time limit a statistically stationary state is reached. For a generic simple fluid, this state would be thermal equilibrium. But the motion of Toda particles is highly constrained through the conservation laws. Still, there are lots and lots of random like collisions. Following Boltzmann, a natural guess for the statistically stationary state is a generalized microcanonical ensemble, namely the uniform measure on the $(N-1)$-dimensional torus $\mathbb{T}^{N-1}$ at fixed values of Lax eigenvalues $\lambda_1, \ldots, \lambda_N$ and of $Q^{[0],N}$, see the discussion in the beginning of Section 2.2. The corresponding thermodynamics thus depends on $N+1$ extensive parameters. As for simple fluids, the first step toward hydrodynamic equations is a study of such generalized thermodynamics.

3.1 Generalized Gibbs ensembles

For fixed number of lattice sites, the phase space is $(r, p) \in \Gamma_N$ with the *a priori* weight

$$\prod_{j=1}^{N} \mathrm{d}r_j \mathrm{d}p_j \delta \left(\sum_{j=1}^{N} r_j - \ell \right) \tag{3.1}$$

for some $\ell \in \mathbb{R}$, where we included already the microcanonical constraint resulting from the boundary conditions (2.9). This measure is invariant under the flow generated by Eq. (2.5). Since particles are distinguishable, there is no factor of $1/N!$ in front. The remaining conserved fields are taken into account through the grand canonical-type Boltzmann weight

$$\exp\left(-\sum_{n=1}^{N} \mu_n Q^{[n],N}\right). \tag{3.2}$$

Here, $\mu_1, \ldots, \mu_N$ are the intensive parameters. Only the low-order ones have a physical interpretation, specifically $\mu_2 = \frac{1}{2}\beta$ with β the inverse temperature and μ_1 as control parameter for the average total momentum. As common in Statistical Mechanics, we invoke the equivalence of ensembles to lift the delta constraint by the substitution

$$\delta(Q^{[0],N} - \ell) \quad \Rightarrow \quad \exp(-PQ^{[0],N}). \tag{3.3}$$

Such kind of equivalence has been extensively studied in rigorous statistical mechanics and presumably some of the techniques can be used also for the Toda lattice. Along with other items, we have to leave this problem for future studies. For a general anharmonic chain, the physical pressure, $\mathfrak{p}$, is defined as the average force between neighboring particles in thermal equilibrium. Using a simple integration by parts, one obtains the relation $\mathfrak{p} = \beta^{-1}P$. We still refer to P as pressure, since it is the thermodynamic dual of the stretch. To have an integrable Boltzmann weight, $P > 0$ is required. In combination, the *generalized Gibbs ensemble* (GGE) is defined through

$$\prod_{j=1}^{N} \mathrm{d}r_j \mathrm{d}p_j \exp\left(-PQ^{[0],N} - \sum_{n=1}^{N} \mu_n Q^{[n],N}\right), \tag{3.4}$$

which still has to be normalized. The GGE is invariant under the Toda dynamical flow.

To study properties of GGE, the first natural step is to transform (3.4) to Flaschka variables. For conciseness, we introduce

$$V(w) = \sum_{n=1}^{\infty} \mu_n w^n. \tag{3.5}$$

The chemical potentials, μ_n, are assumed to be independent of N. Then, the transformed density reads

$$\exp(-\mathrm{tr}[V(L_N)]) \prod_{j=1}^{N} \mathrm{d}p_j \prod_{j=1}^{N} \mathrm{d}a_j \frac{2}{a_j}(a_j)^{2P}, \tag{3.6}$$

which is defined on the phase space of the Flaschka variables, i.e., on Γ_N°.

♦♦ *Confining potential*: A many-particle system is defined by a particular interaction potential, which throughout will be denoted by $V_\bullet$. For example, V_{to} is the interaction potential of the Toda lattice and V_{li} the δ-potential of the Lieb–Liniger model. The potential (3.5) could be called a generalized chemical potential, a terminology which however would entirely miss the central role of V. We wrote V in terms of a power series. But there is no compelling reason to do so. V is simply a rather generic function on $\mathbb{R}$. In the context of GGE, its purpose is to properly confine the eigenvalues of L_N. We thus assume that V is continuous and bounded linearly from below as $V(w) \geqslant c_0 + c_1|w|$ with $c_1 > 0$. Based on such reasoning, V is called *confining potential*. It carries no relation to the interaction potential. Rather V should be viewed as a thermodynamic variable, which so as to speak labels the GGEs. The quadratic confining potential, $V(w) = \frac{1}{2}\beta w^2$, corresponds to thermal equilibrium. ♦♦

♦♦ *Infinite volume limit, exponential mixing, equivalence of ensembles*: As for other Gibbs measures, one might want to know about the existence of the infinite volume limit for the normalized sequence of measures in (3.6), the limit measure being independent of boundary conditions, and a bound on the decay of correlations. If the confining potential is a finite polynomial with strictly positive even leading term, say κ, then the confining potential is bounded from below and such properties can be answered by using transfer matrix techniques. In the language of statistical mechanics, $Q^{[n]}$ has a range of size n and hence $\mathrm{tr}[V(L_N)]$ has range κ. One cuts $[1, \ldots, N]$ in blocks of size κ. The density in (3.6) can then be written as an (N/κ)-fold power of the transfer matrix. This is just like the familiar case of the one-dimensional Ising model, in which case the transfer matrix is a 2×2 matrix. For the Toda lattice, the transfer matrix is given by an integral kernel with arguments in Γ_κ°. By the Perron–Frobenius theorem, the transfer matrix has a unique maximal eigenvalue, which is separated by a gap from the rest of the spectrum. With this input, one concludes that there is a unique limit measure. In one dimension, phase transitions would occur only if the interaction potential has a decay slower than $(\text{range})^{-2}$, much slower than the case under consideration here. The spectral gap also ensures exponential decay of correlations. If $\kappa = \infty$, such methods fail completely. Other techniques will have to be developed, see Notes.

A presumably more delicate issue is the *equivalence of ensembles*. This refers to replacing the sharp constraint $\delta(H_N - eN)$ by $\exp(-\beta H_N)$. On the thermodynamic level, in the limit $N \to \infty$, this amounts to a

Legendre transform between e and β. As a stronger property, the statistics of local observables is the same provided e and β are related according to thermodynamics. For the Toda lattice, we switched from νN to P, a step which is expected to be accessible to current techniques. More delicate is the closed Toda chain as discussed in Section 2.2. In principle, one should fix all action variables. The integral over the tori is trivial and one is left with $\sum_{j=1}^{N} V(\lambda_j)$. Abstractly, this sum depends only on I, but a more concrete characterization does not seem to be available. Thus, a fully microcanonical approach will not be pursued any further.   ◆◆

3.2 Lax matrix filter and local GGEs

This section is somewhat premature, since so far the only model discussed is the Toda lattice. Still, before entering in computational details, we should explain the physics underlying the notion of local GGEs as the central theme of hydrodynamic scales. Some of the material will be explained in greater detail in the following.

Let us start from the example of a classical ideal gas in one dimension, which consists of many particles moving along straight lines as $q_j(t) = q_j + p_j t$ for initial conditions (q, p). The momenta are conserved and hence independent of time. For general random initial data, one introduces the one-particle distribution function

$$\rho_1(x, t; w) = \left\langle \sum_{j \in \mathbb{Z}} \delta(x - q_j(t)) \delta(w - p_j) \right\rangle_{ini}, \qquad (3.7)$$

average over the initial probability density function. For an ideal gas, $\rho_1(x, t; w) = \rho_1(x - wt, 0; w)$.

For the hydrodynamic scale of the ideal gas, we will use the notation $\rho_Q(x, t; w)$ which superficially looks rather similar to ρ_1. Here, (x, t) refers to a macroscopic spacetime point, which is the center of a microscopic cell of size ℓ containing $N(\ell)$ particles. $N(\ell)/\ell$ is of order 1 and each cell contains a large number of particles. The index Q should be a reminder that $\rho_Q(x, t; w)$ carries the information on the average conserved fields in the considered cell. $\rho_Q(x, t; w)$ has a double meaning. We define the empirical velocity distribution function through

$$\frac{1}{\ell} \sum_{j=1}^{N(\ell)} \delta(w - p_j) \simeq \rho_Q(x, t; w), \qquad (3.8)$$

where the sum is over all particles in the cell centered at (x, t). The left hand side is random, but self-averaging in the sense that fluctuations vanish for large $N(\ell)$. In the limit, $\rho_Q(x, t; w)$ is non-random and its integral with respect to w is the macroscopic density $\bar{\rho}(x, t)$. In addition, $\rho_Q(x, t; w)$ uniquely characterizes the local GGE governing the statistics of particles close to (x, t). For an ideal gas, velocities are independent with common probability density function $\rho_Q(x, t; w)/\bar{\rho}(x, t)$. The positional distribution is independent of velocities, more precisely, a Poisson point process with density $\bar{\rho}(x, t)$, which means that inter-particle distances are independent and exponentially distributed. Note that in the limit of infinite scale separation, the local GGE lives on the entire line and is translation invariant. The construction (3.8) holds also for a hard rod fluid. Only now the local GGE has a positional distribution of particles satisfying the hard core constraint.

The at first sight truly surprising claim is that for classical integrable many-body systems the definition in (3.8) can still be used provided the momenta p_j are replaced by the eigenvalues λ_j of the local Lax matrix $L(x, t)$. This is what we mean by a *Lax matrix filter*. Local positions and momenta are noisy. But by inserting these data in the local Lax matrix and determining the local *density of states* (DOS), magically, one filters the slowly varying degrees of freedom. Compared with the ideal gas, rather than sampling local velocities, one has to sample the eigenvalues of the local Lax matrix. More precisely, the Lax matrix is random under the local GGE and has an *empirical* DOS according to

$$\frac{1}{\ell} \sum_{j=1}^{N(\ell)} \delta(w - \lambda_j). \tag{3.9}$$

The sum is over eigenvalues of the Lax matrix constructed from the particular fluid cell under consideration. As valid in great generality, the DOS is self-averaging and thus has a deterministic limit, as before denoted by $\rho_Q(x, t; w)$ but now referring to the Lax DOS. The function $\rho_Q(x, t; w)$ uniquely determines the underlying local GGE in the cell (x, t). The sought for hydrodynamic equations are an evolution equation for $\rho_Q(x, t; w)$, which is of conservation type

$$\partial_t \rho_Q(x, t; w) + \partial_x \rho_J(x, t; w) = 0. \tag{3.10}$$

A major task will be to figure out the $\rho_J(x, t; w)$ as a functional of $\rho_Q(x, t; w)$ at fixed fluid cell (x, t). Concrete examples will come. For some models, the

Lax matrix is unitary and eigenvalues lie on the unit circle. Only through a more detailed analysis such properties can be figured out.

3.3 Generalized free energy

The Toda partition function is defined by

$$Z_{\mathrm{to},N}(P,V) = \int_{\Gamma_N^{\circ}} \prod_{j=1}^{N} \mathrm{d}p_j \prod_{j=1}^{N} \mathrm{d}a_j \frac{2}{a_j} (a_j)^{2P} \exp\big(-\mathrm{tr}[V(L_N)]\big). \qquad (3.11)$$

Accordingly, normalizing the expression in (3.6), one arrives at the probability measure

$$\mu_{P,V,N}^{\mathrm{GGE}}(\mathrm{d}^N a\, \mathrm{d}^N p) = \frac{1}{Z_{\mathrm{to},N}(P,V)} \prod_{j=1}^{N} \mathrm{d}p_j \prod_{j=1}^{N} \mathrm{d}a_j \frac{2}{a_j} (a_j)^{2P} \exp\big(-\mathrm{tr}[V(L_N)]\big). \qquad (3.12)$$

This measure is time-stationary under the dynamics (2.5), since the *a priori* measure and the eigenvalues of L_N do not change in time. The expectations of $\mu_{P,V,N}^{\mathrm{GGE}}$ will be denoted by $\langle\cdot\rangle_{P,V,N}$. As central thermodynamic object, the free energy per lattice site is defined by

$$F_{\mathrm{to}}(P,V) = -\lim_{N\to\infty} \frac{1}{N} \log Z_{\mathrm{to},N}(P,V). \qquad (3.13)$$

For hydrodynamics, a crucial input is the GGE average of the conserved fields. They can be computed as derivative with respect to P and as variational derivative with respect to V of the free energy. In terms of the eigenvalues of the Lax matrix, the averages can be written as

$$\frac{1}{N}\langle Q^{[n],N}\rangle_{P,V,N} = \frac{1}{N}\Big\langle \sum_{j=1}^{N} (\lambda_j)^n \Big\rangle_{P,V,N} = \int_{\mathbb{R}} \mathrm{d}w\, w^n \langle \rho_{\mathrm{Q},N}(w)\rangle_{P,V,N}, \qquad (3.14)$$

where

$$\rho_{\mathrm{Q},N}(w) = \frac{1}{N} \sum_{j=1}^{N} \delta(w - \lambda_j) \qquad (3.15)$$

is the empirical density of states of the Lax matrix L_N. Here, empirical refers to the fact that the DOS is defined for every collection of eigenvalues $\{\lambda_1, \ldots, \lambda_N\}$. As a consequence, $\rho_{\mathrm{Q},N}(w)$ is a random function under $\mu_{P,V,N}^{\mathrm{GGE}}$. More precisely, $\rho_{\mathrm{Q},N}$ is a random probability measure on $\mathbb{R}$ which is supported on N points each with weight $1/N$.

This observation suggests a novel perspective. The Lax matrix becomes a random matrix under $\mu_{P,V,N}^{GGE}$. Thermal equilibrium is particularly simple. Since $\mathrm{tr}[(L_N)^2] = \sum_{j=1}^{N}(p_j^2 + 2a_j^2)$, the diagonal and off-diagonal matrix elements of L_N are families of independent identically distributed (i.i.d.) random variables. For all other GGEs, the matrix elements are correlated. The DOS encodes the complete statistical information on the conserved fields. In fact, $\rho_{Q,N}(w)$ is self-averaging with fluctuations of order $1/\sqrt{N}$ and the limit

$$\lim_{N\to\infty} \rho_{Q,N}(w) = \rho_Q(w) \tag{3.16}$$

exists with some non-random limiting density ρ_Q, which generically is a smooth function.

◆◆ *The Dumitriu–Edelman identity:* In 2002, Dumitriu and Edelman studied the β-ensembles of random matrix theory. To ease a comparison, I describe their result in the original notation, in which some symbols will reappear with a different meaning. The proper translation will be obvious, however. Their starting point is a $n \times n$, symmetric, tridiagonal matrix T with real matrix elements $T_{j,j} = a_j$, $T_{j,j+1} = T_{j+1,j} = b_j > 0$, and zero otherwise, compare with L° from (2.63), in particular, $(a,b) \in \Gamma_n^\circ = \mathbb{R}^n \times \mathbb{R}_+^{n-1}$. The eigenvalues of T are ordered as $\lambda_1 < \ldots < \lambda_n$ and the first component of an eigenvector is denoted by $\psi_j(1) = q_j > 0$ with $|q| = 1$ imposed. We set $d^n a = \prod_{j=1}^{n} da_j$, $d^n\lambda = \prod_{j=1}^{n} d\lambda_j$, and $d^n q$ the surface element of the unit n-sphere. As a general fact of Jacobi matrices with positive off-diagonal matrix elements, there is a one-to-one and onto map $\Psi : (a,b) \mapsto (\lambda,q)$.

Dumitriu and Edelman managed to obtain some information on the Jacobian of Ψ. Their identity has a free parameter $\beta > 0$ and reads

$$\exp\left(-\mathrm{tr}[V(T)]\right) d^n a \prod_{j=1}^{n-1} \frac{2}{b_j}(b_j)^{\beta j} db_j$$

$$= \exp\left(-\sum_{j=1}^{n} V(\lambda_j)\right) \left(n!\tilde{\zeta}_n(\beta)\Delta(\lambda)^\beta d^n\lambda\right) \left(c_q^\beta \prod_{j=1}^{n}(q_j)^{\beta-1} d^n q\right). \tag{3.17}$$

Here,

$$c_q^\beta = 2^{n-1}\Gamma(\tfrac{1}{2}\beta n)\Gamma(\tfrac{1}{2}\beta)^{-n} \tag{3.18}$$

normalizes the third bracket on the right to the integral being 1. In the second bracket, $\Delta(\lambda)$ is the Vandermonde determinant,

$$\Delta(\lambda) = \prod_{1 \leqslant i < j \leqslant n} (\lambda_j - \lambda_i), \tag{3.19}$$

while

$$\tilde{\zeta}_n(\beta) = \Gamma(\tfrac{1}{2}\beta n)^{-1}\Gamma(1 + \tfrac{1}{2}\beta)^n \prod_{j=1}^{n} \frac{\Gamma(\tfrac{1}{2}\beta j)}{\Gamma(1 + \tfrac{1}{2}\beta j)} \tag{3.20}$$

is the overall proportionality constant. ◆◆

In the Dumitriu–Edelman identity (3.17), we substitute $n \rightsquigarrow N$, $a_j \rightsquigarrow p_j$, and $b_j \rightsquigarrow a_j$. Since the scale parameter mentioned below Eq. (2.4) has been set to $\tau = 1$, one notes that $T = L_N^\diamond$. Therefore, integrating both sides of (3.17) over the entire phase space, first making the choice

$$\beta = \frac{2P}{N}, \tag{3.21}$$

yields

$$Z_{\mathrm{de},N}(P,V) = \int_{\Gamma_N^\diamond} \exp\left(-\,\mathrm{tr}[V(L_N^\diamond)]\right) \prod_{j=1}^{N} \mathrm{d}p_j \prod_{j=1}^{N-1} \mathrm{d}a_j \frac{2}{a_j}(a_j)^{2(j/N)P}$$

$$= \zeta_N(P) \int_{\mathbb{R}^N} \mathrm{d}^N\lambda \exp\left(-\sum_{j=1}^{N} V(\lambda_j) + \frac{P}{N} \sum_{i,j=1, i \neq j}^{N} \log|\lambda_i - \lambda_j|\right) \tag{3.22}$$

with prefactor

$$\tilde{\zeta}_N(2P/N) = \zeta_N(P) = \Gamma(P)^{-1}\Gamma(1 + \tfrac{P}{N})^N \prod_{j=1}^{N} \frac{\Gamma(\tfrac{j}{N})}{\Gamma(1 + \tfrac{j}{N})}. \tag{3.23}$$

The integration over eigenvalues is not ordered, which takes care of the factor $N!$. Also, the integration over the N-sphere is normalized to 1.

The term on the right side of (3.22) requires more explanations. For the normalization, one obtains

$$\lim_{N \to \infty} -\frac{1}{N} \log \zeta_N(P) = \log P - 1. \tag{3.24}$$

Otherwise, the partition function is one of the *repulsive one-dimensional log gas*. V turns out to be the confining potential of the log gas, which is the real reason for our original choice of name. In the standard log gas, the interaction strength is of order 1, which implies that the free energy is dominated by the energy term. But in our case, the interaction strength is $1/N$, which is the standard mean-field scaling. Such a problem can be handled through the study of a free energy functional. To distinguish from (3.15), one introduces

$$\varrho_N(w) = \frac{1}{N} \sum_{j=1}^{N} \delta(w - \lambda_j) \tag{3.25}$$

with λ now referring to (3.22). Except for the diagonal contribution, the integrand of (3.22) can be written as

$$\exp\left[-N\left(\int_{\mathbb{R}} dw \varrho_N(w)V(w) - P\int_{\mathbb{R}} dw \int_{\mathbb{R}} dw' \varrho_N(w)\varrho_N(w') \log|w - w'|\right)\right]. \tag{3.26}$$

For given ϱ_N, the corresponding volume element of $d^N\lambda$ is approximately

$$\exp\left[-N\int_{\mathbb{R}} dw \varrho_N(w) \log \varrho_N(w)\right]. \tag{3.27}$$

The large N limit of the partition function is then determined by the mean-field free energy functional

$$\mathcal{F}_{\mathrm{de}}^{\circ}(\varrho) = \int_{\mathbb{R}} dw \varrho(w) \left(V(w) + \log \varrho(w)\right.$$
$$\left. + \log P - P\int_{\mathbb{R}} dw' \varrho(w') \log|w - w'|\right). \tag{3.28}$$

The actual free energy is obtained by minimizing over all ϱ with $\varrho \geqslant 0$ and $\int_{\mathbb{R}} dw \varrho(w) = 1$. As is discussed in the following, there is a unique minimizer, $\varrho^\star$, and thus

$$\lim_{N\to\infty} -\frac{1}{N} \log Z_{\mathrm{de},N}(P,V) = F_{\mathrm{de}}(P,V) = \mathcal{F}_{\mathrm{de}}^{\circ}(\varrho^\star) - 1. \tag{3.29}$$

To obtain the Toda free energy, we note that the Dumitriu–Edelman partition function has a pressure changing linearly with slope $2P/N$. This is not exactly what is required, since the pressure is constant for the Toda chain. But in a large segment ℓ, still with size $\ell \ll N$, the pressure is constant to a very good approximation. Since GGEs have good spatial mixing properties, local free energies merely add up and for the Dumitriu–Edelman free energy, one concludes

$$F_{\mathrm{de}}(P,V) = \int_0^1 du F_{\mathrm{to}}(uP,V) \tag{3.30}$$

and hence

$$F_{\text{to}}(P, V) = \partial_P(PF_{\text{de}}(P, V)). \tag{3.31}$$

While the just presented derivation of $F_{\text{de}}(P,V)$ is fairly standard, one may wonder about the missing steps. A poor man's version will be explained in Section 3.5. Besides, the topic has been studied extensively with methods covering the case of interest. Less standard is the linear pressure ramp leading to the identity (3.30).

In the free energy functional (3.28), the quadratic term has the kernel $\log|w-w'|$ that resulted from the Dumitriu–Edelman change of coordinates. The same expression came up already in Section 2.3 in the context of scattering theory, which might be considered as purely accidental. In fact, we touched upon a generic feature of generalized free energies for integrable models. For the Toda lattice, the connection to scattering theory will be elucidated in Section 9.2.

It turns out to be more convenient to absorb P into ϱ by setting $\rho = P\varrho$. Then, $P\mathcal{F}_{\text{de}}^{\circ}(P^{-1}\rho) = \mathcal{F}_{\text{de}}(\rho)$ with the transformed free energy functional

$$\mathcal{F}_{\text{de}}(\rho) = \int_{\mathbb{R}} \mathrm{d}w \rho(w) \left(V(w) + \log \rho(w) - \int_{\mathbb{R}} \mathrm{d}w' \rho(w') \log|w - w'| \right). \tag{3.32}$$

$\mathcal{F}_{\text{de}}$ has to be minimized under the constraint

$$\rho(w) \geqslant 0, \quad \int_{\mathbb{R}} \mathrm{d}w \rho(w) = P \tag{3.33}$$

with minimizer denoted by $\rho^\star$. Then,

$$F_{\text{to}}(P, V) = \partial_P \mathcal{F}_{\text{de}}(\rho^\star) - 1. \tag{3.34}$$

The constraint (3.33) is removed by introducing the Lagrange multiplier μ as

$$\mathcal{F}_{\text{de}}^{\bullet}(\rho) = \mathcal{F}_{\text{de}}(\rho) - \mu \int_{\mathbb{R}} \mathrm{d}w \rho(w). \tag{3.35}$$

A minimizer of $\mathcal{F}_{\text{de}}^{\bullet}(\rho)$ is denoted by ρ_{n} and by $\rho_{\text{n},\mu}$ when keeping track of the μ-dependence. It is determined as a solution of the Euler–Lagrange equation

$$V(w) + \log \rho_{\text{n},\mu}(w) - \mu - 2 \int_{\mathbb{R}} \mathrm{d}w' \rho_{\text{n},\mu}(w') \log|w - w'| = 0. \tag{3.36}$$

The Lagrange parameter μ has to be adjusted such that

$$P = \int_{\mathbb{R}} \mathrm{d}w \rho_{\text{n},\mu}(w). \tag{3.37}$$

In fact, it will be more convenient to work directly with ρ_n. Note that the Lagrange parameter μ amounts to shifting the confining potential as $V - \mu$. Thus, μ could be incorporated in the definition (3.5) of V as $\mu_0 = -\mu$. The minus sign is a standard convention for the chemical potential dual to the particle number.

To obtain the Toda free energy, we differentiate as

$$\partial_P \mathcal{F}_{\text{de}}(\rho^\star) = \int_{\mathbb{R}} dw \partial_P \rho^\star(w) \left(V(w) + \log \rho^\star(w) \right.$$

$$\left. -2 \int_{\mathbb{R}} dw' \rho^\star(w') \log |w - w'| \right) + 1. \tag{3.38}$$

Integrating (3.36) against $\partial_P \rho^\star$, one arrives at

$$\partial_P \mathcal{F}_{\text{de}}(\rho_{n,\mu}) = \mu + 1 \tag{3.39}$$

and thus

$$F_{\text{to}}(P, V) = \mu(P, V). \tag{3.40}$$

Sharing with other integrable models, the Toda lattice has the property that its free energy is determined by a variational problem for densities over $\mathbb{R}$, in our case normalized to P.

♦♦ *Densities*: While we encountered already the DOS densities ρ_Q and ρ_J, there will be more densities showing up. One has to carefully distinguish between ρ and ϱ. We use ϱ for normalized densities, $\int dw \varrho(w) = 1$, while ρ has a context dependent normalization. For Toda and other integrable systems, the densities are defined on $\mathbb{R}$. Later on, we will also encounter models where the densities live on the unit circle.

To be clear, in our context, density does not refer to position space. The appropriate picture is a density in distorted momentum space. Since there are several densities, one has to distinguish them by a label which in our notation appears through a lower index, as in ρ_Q and ρ_J. We introduced already the density ρ_n, also called *number density*. But it will turn out to be convenient to introduce the further densities ρ_p, ρ_h, and ρ_s called *particle, hole,* and *space densities*. The label is now in serif to avoid confusion with arguments. All densities are functionals of V, a dependence which is mostly suppressed in our notation. For example, ρ_n functionally depends on $V - \mu$, hence it is a function of μ and written as $\rho_{n,\mu}$. ρ_n can be regarded also as a function of P, since $\mu = \mu(P)$ for fixed V. ♦♦

3.4   Lax density of states, TBA equation

To obtain the hydrodynamic equations, GGE averages of the conserved fields are required. As common practice in statistical mechanics, they are defined in the infinite volume limit. For this purpose, we adopt the volume $[-N, \ldots, N]$ with periodic boundary conditions and adjust our notation by using the label $2N + 1$ instead of N. Then, using translation invariance, in the infinite volume limit,

$$\lim_{N \to \infty} \frac{1}{2N+1} \langle Q^{[n],2N+1} \rangle_{P,V,2N+1}$$
$$= \lim_{N \to \infty} \langle Q_0^{[n],2N+1} \rangle_{P,V,2N+1} = \langle Q_0^{[n]} \rangle_{P,V}. \qquad (3.41)$$

On the right-hand side, $Q_0^{[n]}$ is a local function and $\langle \cdot \rangle_{P,V}$ refers to the average with respect to the infinite volume GGE at parameters P, V. Since in our context boundary terms should be negligible, we assume this measure to be well defined and independent of boundary conditions. More pragmatically, only for particular observables, as $Q_0^{[n]}$ and $Q_0^{[n]} Q_j^{[m]}$, the infinite volume average has to exist.

To determine $\langle Q_0^{[n]} \rangle_{P,V}$, one can start from the microscopic definition above and use that $Q^{[n],N}$ depends only on the eigenvalues of the Lax matrix. The other method, employed here, is to simply differentiate the infinite volume free energy. We start with $n = 0$ and note that the average stretch

$$\nu = \langle Q_0^{[0]} \rangle_{P,V} = \partial_P F_{\text{to}}(P,V) = \partial_P \mu(P,V) = \left(\int_{\mathbb{R}} \mathrm{d}w \partial_\mu \rho_{\text{n}}(w) \right)^{-1}, \quad (3.42)$$

where the last equality results from differentiating Eq. (3.37) as

$$1 = \partial_P \mu(P) \int_{\mathbb{R}} \mathrm{d}w \partial_\mu \rho_{\text{n},\mu}(w). \qquad (3.43)$$

For $n \geq 1$, we perturb V as $V_\kappa(w) = V(w) + \kappa w^n$ and differentiate the free energy at $\kappa = 0$. Then,

$$\langle Q_0^{[n]} \rangle_{P,V} = \partial_\kappa F_{\text{to}}(P,V_\kappa)\big|_{\kappa=0} = \partial_P \partial_\kappa \mathcal{F}_{\text{de}}(\rho^\star(P,V_\kappa))\big|_{\kappa=0} \qquad (3.44)$$

and, first introducing the linearization of $\rho^\star$ as

$$\rho^{\star\prime} = \partial_\kappa \rho^\star(P,V_\kappa)\big|_{\kappa=0}, \qquad (3.45)$$

one obtains

$$\partial_\kappa \mathcal{F}_{\text{de}}(\rho^\star(P,V_\kappa))\big|_{\kappa=0} = \int_{\mathbb{R}} \mathrm{d}w \rho^\star(w,P,V) w^n + \int_{\mathbb{R}} \mathrm{d}w \rho^{\star\prime}(w)$$

$$\times \left(V(w) + \log \rho^\star(w,P,V) - 2 \int_{\mathbb{R}} \mathrm{d}w' \rho^\star(w',P,V) \log|w - w'| \right) \qquad (3.46)$$

using that $\int dw \rho^{\star\prime}(w) = 0$. Integrating the Euler–Lagrange equation (3.36) at $\mu = \mu(P)$ against $\rho^{\star\prime}$, the terms on the right-hand side of (3.45) vanish and

$$\langle Q_0^{[n]} \rangle_{P,V} = \int_{\mathbb{R}} dw \partial_P \rho^{\star}(w, P, V) w^n. \tag{3.47}$$

Thus, the Lax DOS is given by

$$\rho_Q(w) = \partial_P \rho^{\star}(w). \tag{3.48}$$

Naively, one might have guessed that the Lax DOS equals $\varrho^{\star}$. But the slow linear variation of the pressure in the Dumitriu–Edelman identity amounts to a slightly deviating result.

In the literature, the Euler–Lagrange equation (3.36) is written differently by formally defining a Boltzmann weight as

$$\rho_{\mathsf{n}}(w) = \mathrm{e}^{-\varepsilon(w)} \tag{3.49}$$

with quasienergy $\varepsilon(w)$. Then,

$$\varepsilon(w) = V(w) - \mu - 2 \int_{\mathbb{R}} dw' \log|w - w'| \mathrm{e}^{-\varepsilon(w')}. \tag{3.50}$$

In addition, one also introduces the particle density through

$$\rho_{\mathsf{p},\mu}(w) = \partial_\mu \rho_{\mathsf{n},\mu}(w), \tag{3.51}$$

at the moment just a convenient terminology.

The structure uncovered is familiar from the Yang–Yang thermodynamics of the Lieb–Liniger δ-Bose gas, which is an integrable quantum many-body system and solved by Bethe ansatz. For quantum integrable systems, the analog of (3.50) is called TBA (thermodynamic Bethe ansatz) equation. We will call (3.50) *classical TBA equation* or simply TBA despite the fact that no Bethe ansatz had been used in its derivation. Some patience is required to fully appreciate this analogy. Further evidence will be accumulated from an alternative route based on scattering coordinates, see Section 9.2 for the Toda lattice and Section 11.3 for the Calogero fluid. The quantum side of the analogy will be covered in Chapter 13 for the δ-Bose gas and in Chapter 14 for the quantum Toda lattice. The exact correspondence will be discussed in Section 13.3.

Later on, we will use some identities based on TBA. We collect them here, together with introducing standard notations. The Hilbert space of square integrable functions on the real line is denoted by $L^2(\mathbb{R}, dw)$ with scalar product

$$\langle f, g \rangle = \int_{\mathbb{R}} dw \bar{f}(w) g(w). \tag{3.52}$$

We will work mostly with real functions and then the complex conjugation in (3.52) can be omitted. There will be many integrals over $\mathbb{R}$ and a convenient shorthand is simply

$$\langle f \rangle = \langle 1, f \rangle = \int_{\mathbb{R}} \mathrm{d}w f(w). \tag{3.53}$$

To distinguish, an average over some probability measure is denoted by $\langle \cdot \rangle_{P,V,N}$, carrying suitable subscripts.

⬩⬩ *Brackets*: Since our convention deviates from widely spread usage in statistical mechanics, to repeat

$\langle h \rangle = \int_{\mathbb{R}} \mathrm{d}w h(w)$ for some function h on $\mathbb{R}$,

$\langle H \rangle_{\bullet,\bullet}$ = average of some random variable H with the probability measure indexed by $\bullet, \bullet$. ⬩⬩

Starting from the $Q^{[n]}$s, so far a discrete basis has been used. Obviously, any linear combination of conserved fields is still conserved and, as in other linear problems, the choice of basis is an important consideration. From the viewpoint of Lax DOS, the label n corresponds to the monomial w^n, which will continued to be used and is denoted by

$$\varsigma_n(w) = w^n, \tag{3.54}$$

including $\varsigma_0(w) = 1$. More generally, the set of basis functions will depend on the particular integrable model under consideration.

Let us define the integral operator

$$Tf(w) = \int_{\mathbb{R}} \mathrm{d}w' \phi_{\mathrm{to}}(w - w')f(w') = 2 \int_{\mathbb{R}} \mathrm{d}w' \log|w - w'|f(w') \tag{3.55}$$

with $w \in \mathbb{R}$. Then, the TBA equation can be rewritten as

$$\varepsilon(w) = V(w) - \mu - (Te^{-\varepsilon})(w). \tag{3.56}$$

In addition, one introduces the dressing of a real-valued function f through

$$f^{\mathrm{dr}} = f + T\rho_{\mathrm{n}} f^{\mathrm{dr}}, \quad f^{\mathrm{dr}} = (1 - T\rho_{\mathrm{n}})^{-1} f, \tag{3.57}$$

where ρ_{n} is regarded as multiplication operator, i.e., $(\rho_{\mathrm{n}} f)(w) = \rho_{\mathrm{n}}(w)f(w)$. With our improved notation, the Lax DOS (3.48) can be written as

$$\rho_{\mathrm{Q}} = \partial_P \rho_{\mathrm{n}} = (\partial_P \mu) \partial_\mu \rho_{\mathrm{n}} = \nu \rho_{\mathrm{p}}, \tag{3.58}$$

where (3.51) has been used. Since ρ_{Q} is normalized,

$$\nu \langle \rho_{\mathrm{p}} \rangle = 1. \tag{3.59}$$

Physically, the central objects of the theory are ν and $\nu \rho_{\mathrm{p}}$, since they encode the GGE average of the conserved fields. Differentiating TBA with respect to μ, we conclude

$$\rho_{\mathsf{p}} = (1 - \rho_{\mathsf{n}} T)^{-1} \rho_{\mathsf{n}} = \rho_{\mathsf{n}} (1 - T\rho_{\mathsf{n}})^{-1} \varsigma_0 = \rho_{\mathsf{n}} \varsigma_0^{\mathrm{dr}}, \qquad (3.60)$$

which expresses ρ_{p} as a functional of ρ_{n}. This relation can be inverted to yield

$$\rho_{\mathsf{n}}(w) = \frac{\rho_{\mathsf{p}}(w)}{1 + T\rho_{\mathsf{p}}(w)}. \qquad (3.61)$$

For later purposes, we also state the definition

$$q_n = \langle Q_0^{[n]} \rangle_{P,V} = \nu \langle \rho_{\mathsf{p}} \varsigma_n \rangle, \qquad (3.62)$$

$n \geqslant 1$, not to be confused with a particle position.

Identities as (3.56) to (3.61) will reappear in other, either classical or quantum, integrable systems. Due to their wide use, these identities are referred to as *TBA formalism* which in a specific way reflects the underlying free energy functional.

◆◆ *TBA equation, uniqueness of solutions*: For the Toda lattice at given μ, the TBA equation has two solutions. At first glance, this looks surprising. In fact, in a standard numerical solution scheme, one follows a particular branch, say starting from small P, and encounters an end point at which instabilities arise. So, some explanations are in demand.

First, $\mu(P)$ is concave and its derivative, $\nu(P)$, is strictly decreasing. For example, in the case of thermal equilibrium, $\mu(P) = \log \sqrt{\beta/2\pi} + P \log \beta - \log \Gamma(P)$, which has a single maximum at $P = P_{\mathrm{c}}$. The physics is rather obvious. At very small P, the average stretch is huge and diverges as $P \to 0$. By increasing pressure, the stretch is decreased. Since there is no hard core, increasing P even further the stretch becomes negative. In physical space, for small P, up to small random errors, the labeling of particles is increasing. But at large P, the labeling is reversed. At P_{c}, the stretch vanishes and the typical distance between particles with adjacent index is of order $1/\sqrt{N}$. The function inverse to $\mu(P)$ has two branches, meaning that for given μ, there are two values of P.

Now considering the densities, by construction $\rho_{\mathsf{n}} \geqslant 0$, $\nu\rho_{\mathsf{p}} \geqslant 0$, $\nu\langle\rho_{\mathsf{p}}\rangle = 1$. $\rho_{\mathsf{n},\mu(P)}$ is pointwise increasing in P and varies smoothly through P_{c}, so does $\nu\rho_{\mathsf{p},\mu(P)}$. On the other hand, $\rho_{\mathsf{p},\mu(P)}(w)$ diverges to $+\infty$ as P approaches P_{c} from the left, globally flips to $-\infty$ at P_{c}, and then flattens out as $P \to \infty$. ◆◆

3.4.1 *Thermal equilibrium*

Thermal equilibrium corresponds to the quadratic confining potential $V(w) = \frac{1}{2}\beta w^2$ with β the inverse temperature. Only for this particular case the diagonal entries of the Lax matrix, $\{p_j, j \in \mathbb{Z}\}$, are independent with p_j a Gaussian random variable of mean zero and variance β^{-1}. Hence, $\langle (p_j)^n \rangle_{P,\beta} = 0$ for odd n and $\langle (p_j)^n \rangle_{P,\beta} = (n-1)!!(\beta)^{-n/2}$ for even n. The off-diagonal entries, $\{a_j, j \in \mathbb{Z}\}$, are also independent with a_j a χ distributed random variable with parameter $2P$. In particular, for the even moments, $\langle (a_j)^{2n} \rangle_{P,\beta} = P(P+1)\ldots(P+n-1)$, $n = 1, 2, \ldots$. Due to independence, the free energy of the chain is easily computed with the result

$$F_{\text{eq}}(P, \beta) = \log\sqrt{\beta/2\pi} + P\log\beta - \log\Gamma(P). \qquad (3.63)$$

However, to figure out the entire DOS requires the TBA machinery.

We start from the Euler–Lagrange equation for the free energy (3.28), set $V(w) = \frac{1}{2}\beta w^2$, differentiate with respect to w, and multiply the resulting expression by $\varrho^\star$. Then,

$$(\beta w + \partial_w)\varrho^\star(w) - 2P \int_{\mathbb{R}} \mathrm{d}w' \frac{1}{w - w'} \varrho^\star(w)\varrho^\star(w') = 0. \qquad (3.64)$$

Note that β scales by setting

$$\varrho^\star_{P,\beta}(w) = \sqrt{\beta}\varrho^\star_{P,1}(\sqrt{\beta}w). \qquad (3.65)$$

Hence, for simplicity, we set $\beta = 1$, omit the explicit dependence on P, and denote by $\varrho^\star$ the solution to (3.64).

Taking the Stieltjes transform,

$$g(z) = \int_{\mathbb{R}} \mathrm{d}w\varrho^\star(w)\frac{1}{w - z}, \qquad (3.66)$$

yields the equation

$$zg(z) + \frac{\mathrm{d}}{\mathrm{d}z}g(z) + Pg(z)^2 = -1. \qquad (3.67)$$

Setting $g(z) = u'(z)/u(z)$, Eq. (3.67) transforms to the linear second-order differential equation

$$u''(z) + zu'(z) + Pu(z) = 0. \qquad (3.68)$$

Since $\varrho^\star$ is a probability density, the $|z| \to \infty$ asymptotics,

$$u(z) \simeq \frac{c_1}{z^P}, \qquad (3.69)$$

follows. Finally, changing to the function $u(z) = e^{z^2/4}y(z)$, one arrives at the Schrödinger-type equation

$$y''(z) + \left(P - \tfrac{1}{2} - \tfrac{1}{4}z^2\right)y(z) = 0, \tag{3.70}$$

which can be solved in terms of parabolic cylinder functions. The appropriate linear combination is determined by the asymptotic condition (3.69). Somewhat unusually, the Stieltjes transform can be still inverted and yields the fairly explicit expression

$$\varrho^\star(w) = \frac{\Gamma(P)e^{-w^2/2}}{\sqrt{2\pi}P|\hat{D}_P(w)|^2}, \quad \hat{D}_P(w) = \int_0^\infty dt\, t^{P-1}e^{iwt}e^{-\frac{1}{2}t^2}. \tag{3.71}$$

Particular examples are shown in Figure 3.1, which however are obtained from numerically solving the nonlinear Fokker–Planck equation (3.64) rather than using (3.71).

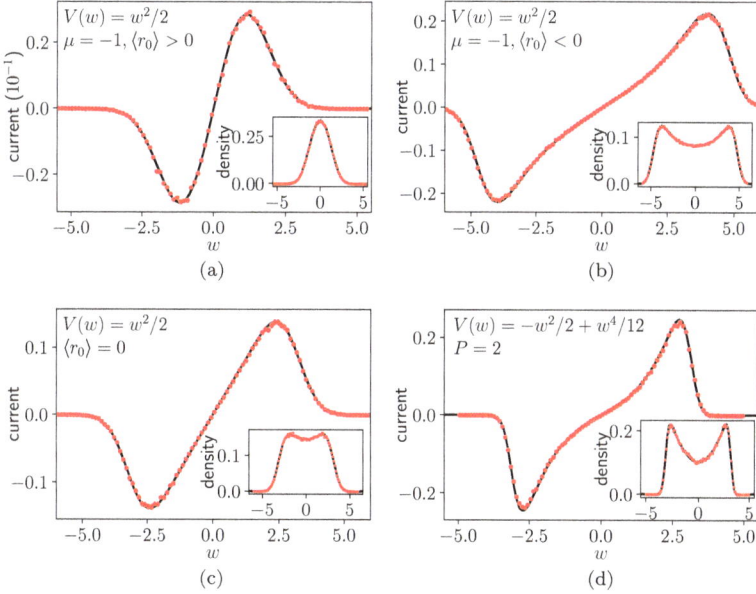

Fig. 3.1. For the Toda lattice shown is the current DOS ρ_J and as inset the corresponding DOS ρ_Q. For panels (a), (b), (c), the Toda lattice is in thermal equilibrium with $\beta = 1$ and $(\nu, P) = (7.64, 0.138)$, $(-1.28, 4.10)$, $(0, 1.461)$, respectively. Panel (c) corresponds to the critical pressure P_c, while (a) and (b) have the same value of μ, thereby confirming that the TBA equation has two distinct solutions. Panel (d) displays the same simulation for a quartic confining potential at $P = 2$, $\beta = 1$. For the DOS, sampled are the eigenvalues of the Lax matrix at $N = 1,024$ (red dots) and compared with the numerical solution of the TBA equation as based on the nonlinear Fokker–Planck equation (4.11). For the current DOS, the expression (2.25) is sampled from the random Lax matrix and compared with ρ_J computed through the TBA formalism, see Eq. (6.18). From Cao *et al.* (2019).

We reintroduce the dependence on P, β. For small P, the Lax off-diagonal matrix elements $a_j \to 0$ and hence in leading order $\varrho^\star_{P,\beta}(w) = \sqrt{\beta/2\pi} \exp(-\frac{1}{2}\beta w^2)$. On the other hand, one can integrate (3.64) against $\varsigma_n(w)$ to obtain a recursion relation for the even moments of $\varrho^\star_{P,\beta}$, $c_n = \langle \varrho^\star_{P,\beta}(w)w^{2n} \rangle$, more concretely

$$\beta c_n = (2n-1)c_{n-1} + \mathfrak{p}\beta \sum_{j=0}^{n-1} c_{n-1-j}c_j, \tag{3.72}$$

with $n = 1, 2, \ldots$ and $c_0 = 1$, where we switched to the physical pressure $\mathfrak{p} = P/\beta$. For low temperatures, $\beta \to \infty$, the term $(2n-1)c_{n-1}(\beta)$ can be neglected implying the asymptotic result $c_n(\beta) \to c_n(\infty)$ with

$$c_n(\infty) = \mathfrak{p}^n \frac{1}{n+1} \binom{2n}{n}. \tag{3.73}$$

On the right-hand side, one notes the Catalan numbers. Hence, $\varrho^\star_{\mathfrak{p}\infty,\infty}$ is the normalized Wigner semi-circle probability density function

$$\varrho^\star_{\mathfrak{p}\infty,\infty}(w) = \frac{1}{2\pi\mathfrak{p}} \sqrt{4\mathfrak{p} - w^2} \tag{3.74}$$

for $|w| \leqslant 2\sqrt{\mathfrak{p}}$. To obtain the Lax DOS, one still has to act with the operator $\partial_P P$. The low-pressure Gaussian does not change. For low temperatures, upon applying the operator $\partial_\mathfrak{p}\mathfrak{p}$ one concludes the convergence

$$\lim_{\beta\to\infty, P=\mathfrak{p}\beta} \rho_Q(w) = \frac{1}{\pi\sqrt{4\mathfrak{p} - w^2}} \tag{3.75}$$

for $|w| \leqslant 2\sqrt{\mathfrak{p}}$ in the sense of convergence of moments. The Lax eigenvalues concentrate close to the borders $\pm 2\sqrt{\mathfrak{p}}$. In probability theory, the density (3.75) is known as centered arcsine law.

This convergence can also be obtained from the random walk expansion of $\langle Q_0^{[n]} \rangle_{\mathfrak{p}\beta,\beta}$. Then, every path with at least one horizontal step carries a weight with at least one factor of β^{-1}. Thus, in the limit, one has to count only the number of simple random walks with n steps starting at $j = 0$ and ending at $j = 0$, each step carrying the weight $\mathfrak{p}^{1/2}$. This number is non-zero only for even n, in which case $\langle Q_0^{[2n]} \rangle_{\mathfrak{p}\infty,\infty} = (n+1)c_n(\infty)$.

3.5 Mean-field techniques

Not to interrupt the line of arguments, we only claimed that the free energy for the Dumitriu–Edelman partition function can be obtained by minimizing the free energy functional $\mathcal{F}^{\circ}_{\text{de}}$. In fact, in all integrable models studied so far, both classical and quantum, there is such a variational characterization for the generalized free energy. Thus, it is worthwhile to elucidate the main characteristics independently from the specific applications we are interested in. From a probabilistic perspective, the problem belongs to the theory of large deviations, as pioneered by Donsker and Varadhan. For our purposes, a more simple-minded statistical mechanics approach will do. But beware, we only explain the main steps. More space would be needed for a complete proof in particular cases.

We will outline what mathematicians call a proof by compactness. A simple example would be a sequence of real numbers $y_1, y_2, \ldots$ with $|y_j| \leqslant a$, the condition of compactness. One wants to determine the large j limit $y_j \to y_\infty$. Due to uniform bound, one can choose a subsequence $\tilde{y}_j$ such that $\tilde{y}_j \to \tilde{y}_\infty$ for $j \to \infty$. The crucial input is some information on $\tilde{y}_\infty$. In our context a natural example would be to have some function $g : [-a, a] \to \mathbb{R}$ with the property that $g(\tilde{y}_\infty) = 0$. Assuming in addition that g has a unique zero, $\bar{y}$, one concludes that $\lim_{j \to \infty} \tilde{y}_j = \bar{y}$. Since this argument applies to any convergent subsequence, one arrives at $\lim_{j \to \infty} y_j = \bar{y}$. To obtain some information on the rate of convergence would require additional considerations.

We turn to the mean-field-type problem by considering N "spins" $x_1, \ldots, x_N$ with $x_j \in \mathbb{R}$. The "energy" of a spin configuration is taken to be of the particular form

$$H_N(x) = \sum_{j=1}^{N} V(x_j) + \frac{1}{N} \sum_{i,j=1}^{N} W(x_i, x_j). \tag{3.76}$$

Thereby, one defines partition function and Gibbs probability density relative to $\mathrm{d}^N x$,

$$Z_N = \int_{\mathbb{R}^N} \mathrm{d}^N x \exp\big(-H_N(x)\big), \quad \mu^N(x) = \frac{1}{Z_N} \exp\big(-H_N(x)\big). \tag{3.77}$$

The superscript N refers to a probability density on $\mathbb{R}^N$ with standard definitions for free energy, energy, and entropy,

$$F(\mu^N) = -\log Z_N, \quad E(\mu^N) = \langle H_N \rangle_{\mu^N}, \quad S(\mu^N) = -\int_{\mathbb{R}^N} \mathrm{d}^N x \mu^N \log \mu^N. \tag{3.78}$$

As before, the main goal is the large N limit of the empirical density

$$\varrho_N(w) = \frac{1}{N} \sum_{j=1}^{N} \delta(w - x_j). \tag{3.79}$$

The energy of our model is very special in the sense that it is merely a functional of the empirical density,

$$H_N = N \int_{\mathbb{R}} \mathrm{d}w \varrho_N(w) \left(V(w) + \int_{\mathbb{R}} \mathrm{d}w' \varrho_N(w') W(w, w') \right). \qquad (3.80)$$

The quadratic functional serves here only as illustration. More complicated functionals will appear in Chapters 9 and 11. Of course, the potential V has to be sufficiently confining. In a first run, the kernel $W(w, w') = W(w', w)$ could be taken to be bounded. As central observation, μ^N is invariant under permutation of indices. As a striking consequence, for example, the two-point function $\langle f_1(x_i) f_2(x_j) \rangle_{\mu^N}$ does not depend on the choice of (i, j). Clearly, such a large symmetry group will force special properties.

The mentioned compactness can be ensured by bounds as $Z_N \leqslant \mathrm{e}^{c_0 N}$. Thus, there exists a subsequence, denoted again as μ^N, such that $\mu^N \to \mu_\infty$. Clearly, the limit measure μ_∞ is invariant under permutations, a property called exchangeable. The set of all exchangeable measures on $\mathbb{R}^{\mathbb{N}}$ is denoted by $\mathcal{P}$. Now by a theorem of Hewitt and Savage, in a simplified context earlier studied by De Finetti, any exchangeable probability measure must be a convex combination of product probability measures. A pure product measure is written as

$$\mu_\varrho = \varrho \times \varrho \times \cdots \qquad (3.81)$$

with $\varrho \geqslant 0$ and $\langle \varrho \rangle = 1$. Each factor has finite entropy and hence is of the form $\varrho(w) \mathrm{d}w$ with finite

$$S(\varrho) = - \int_{\mathbb{R}} \mathrm{d}w \varrho(w) \log \varrho(w). \qquad (3.82)$$

The entropy and energy per site equal

$$s(\mu_\varrho) = S(\varrho), \quad e(\mu_\varrho) = \langle V, \varrho \rangle + \langle \varrho, W \varrho \rangle = E(\varrho). \qquad (3.83)$$

The single site probability density is assumed to also have finite energy. The set of all measures with finite entropy and energy is denoted by $\mathcal{M}$. A convex combination of product measures, i.e., any $\mu \in \mathcal{P}$, can then be written as

$$\mu = \int_{\mathcal{M}} \nu(\mathrm{d}\varrho | \mu) \mu_\varrho. \qquad (3.84)$$

Entropy and energy per site now turn to

$$s(\mu) = \int_{\mathcal{M}} \nu(\mathrm{d}\varrho | \mu) S(\varrho), \quad e(\mu) = \int_{\mathcal{M}} \nu(\mathrm{d}\varrho | \mu) E(\varrho). \qquad (3.85)$$

While the second identity is obvious, the first one looks wrong and is so at finite volume. But consider $\alpha \varrho_1 \times \cdots \times \varrho_1 + (1 - \alpha) \varrho_2 \times \cdots \times \varrho_2$, each product having N factors. For large N, these densities are supported on essentially disjoint sets in phase space and the integral defining the total entropy is in leading approximation $\alpha N S(\varrho_1) + (1 - \alpha) N S(\varrho_2)$.

We now choose a (sub)sequence such that $\mu^N \to \mu_\infty$. Then, the limit

$$\lim_{N \to \infty} \frac{1}{N} E(\mu^N) = e(\mu_\infty) \tag{3.86}$$

exists.

(i) Lower bound: For μ^N, the marginal onto $[1, \ldots, n]$, $n \leqslant N$, is denoted by μ_n^N and correspondingly by μ_n for some infinite volume measure μ. We consider large N and divide the volume into disjoint segments each of length n. Then, by subadditivity of entropy,

$$S(\mu^N) \leqslant \left\lfloor \frac{N}{n} \right\rfloor S(\mu_n^N) + \mathcal{O}(1) \tag{3.87}$$

with $\lfloor \cdot \rfloor$ indicating integer part. Therefore,

$$\lim_{N \to \infty} \frac{1}{N} S(\mu^N) \leqslant \lim_{n \to \infty} \frac{1}{n} s(\mu_{\infty,n}) = s(\mu_\infty) \tag{3.88}$$

and, since $f = e - s$,

$$\lim_{N \to \infty} \frac{1}{N} F(\mu^N) \geqslant e(\mu_\infty) - s(\mu_\infty) = f(\mu_\infty). \tag{3.89}$$

(ii) Upper bound: By the standard finite volume variational principle for the free energy

$$\lim_{N \to \infty} \frac{1}{N} F(\mu^N) \leqslant \lim_{N \to \infty} \min_{\mu' \in \mathcal{P}} \frac{1}{N} F(\mu'_N) = \min_{\mu' \in \mathcal{P}} f(\mu') \leqslant f(\mu_\infty). \tag{3.90}$$

Combining both bounds

$$\lim_{N \to \infty} \frac{1}{N} F(\mu^N) = f(\mu_\infty) = \min_{\mu' \in \mathcal{P}} \int_{\mathcal{M}} \nu(d\varrho'|\mu') \mathcal{F}(\varrho'), \tag{3.91}$$

where the one-particle free energy functional reads

$$\mathcal{F}(\varrho) = \langle \varrho, V + \log \varrho + W \varrho \rangle. \tag{3.92}$$

In mean-field models from equilibrium statistical mechanics, generically, there are several minimizers, indicating a phase transition in the microscopic system. For integrable many-body systems, uniqueness seems to be the rule. Assuming such property to hold and denoting the unique minimizer by ϱ^*, our formulas simplify drastically,

$$\lim_{N\to\infty} \frac{1}{N} F(\mu^N) = \mathcal{F}(\varrho^*). \tag{3.93}$$

Modulo technical points, this is the limit claimed in (3.29). Furthermore, subsequence can be replaced by sequence and

$$\mu^N \to \mu_{\varrho^*}. \tag{3.94}$$

The density of states, integrated against some test function f, equals $\langle f \varrho_N \rangle$ which is still random. From (3.94), one deduces the limits

$$\lim_{N\to\infty} \langle\langle f \varrho_N \rangle\rangle_{\mu^N} = \langle f \varrho^* \rangle, \quad \lim_{N\to\infty} \langle (\langle f \varrho_N \rangle)^2 \rangle_{\mu^N} = \langle f \varrho^* \rangle^2, \tag{3.95}$$

which establishes the law of large numbers for $\langle f \varrho_N \rangle$, in other words, the almost sure convergence of $\langle f \varrho_N \rangle$ to the deterministic limit $\langle f \varrho^* \rangle$ as $N \to \infty$.

In statistical mechanics, mean-field usually refers to an approximation where a fluctuating field is replaced by its average. In contrast, for integrable many-body systems, the respective mean-field problem is exact and arrives unexpectedly. For the Toda lattice, the transformation of Dumitriu–Edelman accomplishes the deal. As to be discussed, for the Calogero fluid, the proof relies on scattering coordinates and for the Ablowitz–Ladik system, on the transformation linked to CMV matrices. For quantum many-body systems, a mean-field-type problem naturally arrives through the Bethe ansatz, more generally through the asymptotic Bethe ansatz.

Notes and references

Section 3.0

At the age of 24, Ludwig Boltzmann (1868) wrote his fundamental contribution on the microcanonical ensemble. He argued that this ensemble provides the natural description of the long time behavior of a mechanical system. In addition, he uncovered the connection to thermodynamics. Recommended is his very readable letter-style account Boltzmann (1868a). The way how we teach equilibrium statistical mechanics today goes back to the must-read book by J. Williard Gibbs (1902).

Section 3.1

The generalized Gibbs ensemble came naturally into focus when studying the quench of integrable systems starting from a translation invariant state, see the reviews Polkovnikov *et al.* (2011) and Vidmar and Rigol (2016), the special volume edited by Calabrese *et al.* (2016), and the more recent short review by Alba and Calabrese (2017) with further references. In this context, the conserved charges averaged over the initial state determine the parameters of the GGE obtained in the long time limit. In some circles, the notion "generalized" was initially not so welcome, since there is a long tradition in the study of Gibbs measures for a general class of potentials, see Ruelle (1969), Lanford (1973), Simon (2016), and Friedli and Velenik (2017) for a tiny subset. In our context, "generalized" means that for the given mechanical system one has a high dimensional set of time-invariant measures all of them having the form anticipated by Statistical Mechanics.

Equivalence of ensembles in its original meaning refers to the property that the resulting thermodynamic potentials are related to each other through the respective Legendre transform. This is a widely studied subject, with close relations to the theory of large deviations, as discussed by Ruelle (1969), Lanford (1973), Simon (2016), and Friedli and Velenik (2017). In our context, we use a stronger version: In the limit of infinite volume, two distinct ensembles constructed from the same bulk Hamiltonian yield identical distributions of strictly local observables. To put it differently, provided the respective parameters are transformed according to the rules of thermodynamics, correlations of local functions are identical. Such a stricter version generically fails at phase transitions; for a detailed discussion, see Georgii (1994, 1975).

Section 3.2

A general problem of non-equilibrium statistical mechanics is to determine the slowly varying fields. The obvious candidates are the locally conserved fields. In higher dimensions, the equilibrium phase diagram has more structure. For example, gas and fluid phase may coexist. The respective phase boundary is then also a slowly varying field. More generally, all order parameters of the thermodynamic phases have to be included in the list, see Forster (1975). Generalized hydrodynamics (GHD) follows the same approach. For the Toda lattice, the Lax filter is a natural construction, which is applicable to other integrable systems, whenever there is a Lax matrix available.

Section 3.3

The generalized free energy of the Toda lattice has been obtained in Spohn (2019), see also Doyon (2018, 2020) for complimentary discussions. Originally, Dumitriu and Edelman (2002) investigated how the fully occupied GUE random matrix transforms isospectrally to a tridiagonal matrix. For this purpose, they iteratively applied the Householder transform, which is a standard numerical scheme. Once Dumitriu and Edelman had understood how this scheme works for GUE random matrices, they realized that their algebra holds for arbitrary values of the parameter β and not only for $\beta = 2$. In their context, the quadratic confining potential appears naturally. Since the transformation of volume elements is established, the extension to general confining potentials is straightforward. The Dumitriu–Edelman identity has been used to obtain a fairly detailed information on the edge behavior of the general β-ensemble, see Ramirez *et al.* (2011) and Bourgade *et al.* (2014) for further developments. A standard reference on log gases is the monumental volume by Forrester (2010). Variational problems with a logarithmic kernel are investigated by Saff and Totik (1997).

In a recent contribution, Guionnet and Memin (2022) prove the generalized free energy to be given by (3.31) and the almost sure limit of the DOS as stated in (3.16) with limit $\nu \rho_\mathsf{p}$. Their theorem covers confining potentials such that $\lim_{|w| \to \infty} w^{-2n} V(w) = a_{\infty,n} > 0$ for some positive integer n. In particular, V is not required to be given by a convergent power series. From the hydrodynamic perspective, required would be also the inverse operation. More precisely, the solution to the hydrodynamic equations determines locally a particular ρ_n. The corresponding V, as obtained from Eq. (3.50), should be admissible. A further line of research, not yet covered at all, concerns the spatial structure of GGEs. This includes the decay of correlation functions and mixing properties. For such an analysis of use could be Dobrushin's statistical mechanics theory, Dobrushin (1974), developed specifically for one-dimensional systems.

Section 3.4

The particular form of the TBA formalism can be better grasped upon reading Section 13. The double-valuedness of solutions to the TBA equation was pointed out to me by Bulchandani *et al.* (2019) with numerical confirmation in Cao *et al.* (2019). Numerical plots of densities as they vary through P_c can be found in Mendl and Spohn (2022). The TBA equation (3.50) with quadratic confining potential was first obtained by Opper (1985), who investigated the classical limit of the quantum Toda chain. He already obtained the solution (3.71). Our discussion is based on Allez *et al.* (2012), a study of Dyson Brownian motion with weak interaction, see Chapter 4 for details. A related study has been carried out for β-Wishart ensembles by Allez *et al.* (2012a). The recursion relation (3.72) appears already in Duy and Shirai (2015) and Duy (2018).

Section 3.5

Out of the many texts and monographs on the theory of large deviations recommended is the book by Dembo and Zeitouni (2010). The complete proof based on permutation symmetry can be found in Messer and Spohn (1982). A more recent study are the lecture notes by Rougerie (2015) who also presents an anologous discussion of quantum mean-field models. De Finetti's theorems on exchangeable measures are discussed in Aldous (1985). The generality required in our context is covered by Hewitt and Savage (1985).

Chapter 4

Dyson Brownian Motion

In the mid-1960s, Freeman Dyson pioneered the study of random matrices, as initiated by Eugene Wigner and others as a phenomenological model for the complex energy spectra of highly excited nuclei. More specifically, Dyson studied the statistics of eigenvalues for the Gaussian orthogonal, unitary, and symplectic ensembles. For this purpose, he introduced a stochastic dynamics of eigenvalues with the property that its stationary distribution coincides with the distribution of eigenvalues of the respective random matrix ensemble. Such a diffusion process is now called Dyson Brownian motion. In this approach, one studies a physically more intuitive stochastic particle system, which in the long time limit approaches the desired probability density of eigenvalues. Particles move in one dimension, their positions being denoted by $\{x_j(t), j = 1, \ldots, N\}$, and are governed by the coupled stochastic differential equations,

$$\mathrm{d}x_j(t) = -V'(x_j(t))\mathrm{d}t + 2\beta \sum_{i=1, i\neq j}^{N} \frac{1}{x_j(t) - x_i(t)}\mathrm{d}t + \sqrt{2}\mathrm{d}b_j(t), \qquad (4.1)$$

$j = 1, \ldots, N$, with $\beta \geqslant 0$. The particles repel each other through a $1/x$ force and are confined by the external potential V, which in fact will coincide with the generalized chemical potential (3.5) of the Toda lattice — in the first place, the reason for using the same name and symbol. $\{b_j(t), j = 1, \ldots, N\}$ is a collection of independent standard Brownian motions. For later purposes, the generator, $\mathcal{L}_N$, of the coupled diffusion processes is defined as the linear operator

$$\mathcal{L}_N f(x) = \sum_{j=1}^{N} \left(\partial_{x_j} - V'(x_j) + 2\beta \sum_{i=1, i\neq j}^{N} \frac{1}{x_j - x_i} \right) \partial_{x_j} f(x), \qquad (4.2)$$

acting on functions on configuration space, $f : \mathbb{R}^N \to \mathbb{R}$, $x = (x_1, \ldots, x_N)$. For $t \geqslant 0$, the kernel of the semi-group

$$e^{\mathcal{L}_N t}(x, x') d^N x' \tag{4.3}$$

is the transition probability, in other words, the probability density at time t given the initial configuration x.

The reader might suspect a luxury detour. But this is not the case. Dyson Brownian motion is a powerful method to numerically solve the TBA equations of the Toda lattice. Even more importantly, as will be discussed, the GGE averaged currents are linked to eigenvalue fluctuations. To study such properties, Dyson Brownian motion is a convenient tool.

Dyson investigated the particular values $\beta = 1, 2, 4$, which possess a large symmetry group. But recently, there has been much progress also for general β. In our context, we will have to study the case of very small β, specifically $\beta = \alpha/N$. In fact, $\alpha = P$ in the application to the Toda lattice. In (4.1), the drift is the gradient of a potential. Hence, the diffusion process is reversible and its unique stationary measure is given by

$$\frac{1}{Z_{\alpha,V,N}} \exp\left(-\sum_{j=1}^N V(x_j) + \frac{\alpha}{N} \sum_{i,j=1,i\neq j}^N \log|x_i - x_j|\right), \tag{4.4}$$

which agrees with the normalized version of the probability density function appearing in (3.22). Our strategy will be to first study the large N limit of the dynamics and then deduce information on the stationary measure, which is our real interest.

4.1 Macroscopic equation, law of large numbers

We choose some smooth test function f and introduce the empirical density, $\Upsilon_N(x, t)$, through

$$\Upsilon_N(f, t) = \frac{1}{N} \sum_{j=1}^N f(x_j(t)) = \int_{\mathbb{R}} dx \Upsilon_N(x, t) f(x). \tag{4.5}$$

Then,

$$d\Upsilon_N(f, t) = \Upsilon_N(-V'\partial_x f + \partial_x^2 f, t)dt + \alpha \int_{\mathbb{R}} dx \int_{\mathbb{R}} dy \frac{f'(x) - f'(y)}{x - y}$$

$$\times \Upsilon_N(x, t)\Upsilon_N(y, t)dt + \frac{1}{N} \sum_{j=1}^N f'(x_j(t))\sqrt{2}db_j(t). \tag{4.6}$$

When integrated in time, the process averaged square of the noise term becomes

$$\frac{2}{N^2} \sum_{j=1}^{N} \int_0^t ds \mathbb{E}\big(f(x_j(s))^2\big). \tag{4.7}$$

Thus, the noise is of order $1/\sqrt{N}$ and vanishes in the limit $N \to \infty$.

We assume a starting measure such that with probability one the initial empirical density has the limit ϱ_0,

$$\lim_{N \to \infty} \Upsilon_N(f, 0) = \int_{\mathbb{R}} dx \varrho_0(x) f(x). \tag{4.8}$$

In the statistical physics literature, this property is often referred to as "self-averaging", meaning that the limit is non-random and no averaging is required. Since in (4.6) only the drift term survives, one concludes the limit

$$\lim_{N \to \infty} \Upsilon_N(f, t) = \Upsilon(f, t) = \int_{\mathbb{R}} dx \varrho(x, t) f(x) \tag{4.9}$$

for all $t > 0$, again with probability one. The limit density $\varrho(x, t)$ then satisfies

$$\frac{d}{dt} \int_{\mathbb{R}} dx \varrho(x, t) f(x) = \int_{\mathbb{R}} dx \varrho(x, t)(-V' \partial_x f + \partial_x^2 f)(x)$$
$$+ \alpha \int_{\mathbb{R}} dx \int_{\mathbb{R}} dy \frac{f'(x) - f'(y)}{x - y} \varrho(x, t) \varrho(y, t) \tag{4.10}$$

together with the initial data $\varrho_0(x)$. Written pointwise, $\varrho(x, t)$ is governed by the nonlinear Fokker–Planck equation

$$\partial_t \varrho(x, t) = \partial_x \big(V'_{\text{eff}}(x, t) \varrho(x, t) + \partial_x \varrho(x, t)\big). \tag{4.11}$$

The bare confining potential V is modified to an effective potential given by

$$V_{\text{eff}}(x, t) = V(x) - \alpha(T\varrho)(x, t), \tag{4.12}$$

the integral operator T being defined in (3.55).

Our particular interest is the stationary Fokker–Planck equation

$$\partial_x \left(V'(x) \varrho_{\mathsf{s}}(x) - 2\alpha \int_{\mathbb{R}} dy \frac{1}{x - y} \varrho_{\mathsf{s}}(y) \varrho_{\mathsf{s}}(x) + \partial_x \varrho_{\mathsf{s}}(x) \right) = 0 \tag{4.13}$$

which, under the constraint

$$\int_{\mathbb{R}} dx \varrho_{\mathsf{s}}(x) = 1, \tag{4.14}$$

has a unique strictly positive solution ϱ_{s}. Since Dyson Brownian motion is time-reversible, the large round bracket itself has to vanish. We compare to the mean-field free energy functional $\mathcal{F}^{\circ}_{\mathrm{de}}(\varrho)$ in Eq. (3.28), setting $P = \alpha$. Its minimizer $\varrho^{\star}$ satisfies the Euler–Lagrange equation

$$V(x) - \mu - 2\alpha \int_{\mathbb{R}} dx' \log |x - x'| \varrho^{\star}(x') + \log \varrho^{\star}(x) + 1 = 0, \qquad (4.15)$$

where w has been substituted by x. Differentiating with respect to x and then multiplying by $\varrho^{\star}$ yields

$$V'(x)\varrho^{\star}(x) - 2\alpha \int_{\mathbb{R}} dy \frac{1}{x-y} \varrho^{\star}(x)\varrho^{\star}(y) + \partial_x \varrho^{\star}(x) = 0. \qquad (4.16)$$

Thus, we conclude $\varrho_{\mathrm{s}} = \varrho^{\star}$.

The nonlinear Fokker–Planck equation depends smoothly on α and so does ϱ_{s}. The issue of two-valuedness appears only in the TBA equation.

4.2 Fluctuation theory

The nonlinear Fokker–Planck equation is the result of a law of large numbers. Thus, one would expect to have a central limit-type theorem which captures the order $1/\sqrt{N}$ fluctuations. Quite generically, the fluctuations are governed by a linear Langevin equation for the conserved field, in our case, the density field. The drift term is determined by the linearized macroscopic equation. But in addition the dynamics generates an effective noise term, whose precise structure depends on the model.

Our interest is the stationary dynamics, hence the initial state of the N-particle system is the one in Eq. (4.4). To study the fluctuations of the density, it is convenient to introduce the fluctuation field as

$$\phi_N(f,t) = \frac{1}{\sqrt{N}} \sum_{j=1}^{N} \left(f(x_j(t)) - \langle \varrho_{\mathrm{s}} f \rangle \right) = \int_{\mathbb{R}} dx f(x)\phi_N(x,t). \qquad (4.17)$$

As explained in more detail in the following, there is a Gaussian random field $\phi(f,t)$, jointly in f and t, such that in distribution

$$\lim_{N\to\infty} \phi_N(f,t) = \phi(f,t). \qquad (4.18)$$

Note that the limit is still random. The limit field ϕ is governed by the linear Langevin equation

$$\partial_t \phi(x,t) = \partial_x \left(D\phi(x,t) + \sqrt{2\varrho_{\mathrm{s}}(x)} \xi(x,t) \right). \qquad (4.19)$$

Here, $\xi(x,t)$ is the normalized spacetime Gaussian white noise, $\mathbb{E}\left(\xi(x,t)\xi(x',t') \right) = \delta(x - x')\delta(t - t')$ and D is the linear operator

$$D = V'_{\mathrm{eff}} + \partial_x - \alpha\varrho_{\mathrm{s}}T'. \qquad (4.20)$$

Thus, $\partial_x D$ is the Fokker–Planck evolution operator linearized at $\varrho_{\mathfrak{s}}$. The effective potential V_{eff} is still defined as in (4.12) upon substituting $\varrho(x,t)$ by $\varrho_{\mathfrak{s}}(x)$.

The Gaussian process $\phi(x,t)$ is stationary in time, has mean zero, and is uniquely characterized by its covariance $\mathbb{E}\big(\phi(x,t)\phi(x',t')\big)$. Of particular interest for the Toda lattice is the spatial covariance

$$\mathbb{E}\big(\phi(x,0)\phi(x',0)\big) = C^{\sharp}(x,x'). \tag{4.21}$$

As a general property of linear Langevin equations with time-independent coefficients, $C^{\sharp}$ is determined by

$$\langle D^*\partial_x f, C^{\sharp}g\rangle + \langle f, C^{\sharp}D^*\partial_x g\rangle = 2\langle \partial_x f, \varrho_{\mathfrak{s}}\partial_x g\rangle \tag{4.22}$$

with D^* the adjoint operator to D. We claim that, as an operator, the solution is

$$C^{\sharp} = (1 - \alpha\varrho_{\mathfrak{s}}T)^{-1}\varrho_{\mathfrak{s}} - \big\langle(1 - \alpha\varrho_{\mathfrak{s}}T)^{-1}\varrho_{\mathfrak{s}}\big\rangle^{-1}$$
$$\times\big|(1 - \alpha\varrho_{\mathfrak{s}}T)^{-1}\varrho_{\mathfrak{s}}\big\rangle\big\langle(1 - \alpha\varrho_{\mathfrak{s}}T)^{-1}\varrho_{\mathfrak{s}}\big|, \tag{4.23}$$

where for simplicity we use the Dirac notation $|\cdot\rangle\langle\cdot|$ for the one-dimensional projector. The subtracted term ensures that the number of particles does not fluctuate, i.e., $C^{\sharp}\varsigma_0 = 0$.

To confirm the claim, we consider only the left most term of (4.22), the other one following by symmetry, and have to show that

$$\langle\partial_x f, D\varrho_{\mathfrak{s}}(1 - \alpha T\varrho_{\mathfrak{s}})^{-1}g\rangle = \langle\partial_x f, \varrho_{\mathfrak{s}}\partial_x g\rangle. \tag{4.24}$$

Upon replacing g by $(1 - \alpha T\varrho_{\mathfrak{s}})g$, one arrives at

$$\langle\partial_x f, D\varrho_{\mathfrak{s}}g\rangle = \langle\partial_x f, \varrho_{\mathfrak{s}}\partial_x(1 - \alpha T\varrho_{\mathfrak{s}})g\rangle. \tag{4.25}$$

$\varrho_{\mathfrak{s}}$ satisfies the stationary nonlinear Fokker–Planck equation (4.16),

$$\big(V'_{\text{eff}} + \partial_x\big)\varrho_{\mathfrak{s}} = 0. \tag{4.26}$$

When inserted in (4.25), this leads to the condition

$$\varrho_{\mathfrak{s}}\partial_x g - \alpha\varrho_{\mathfrak{s}}T'(\varrho_{\mathfrak{s}}g) = \varrho_{\mathfrak{s}}\partial_x(1 - \alpha T\varrho_{\mathfrak{s}})g, \tag{4.27}$$

which is easily checked.

⬧⬧ *Martingales and central limit theorem*: In probability theory, tremendous efforts have been invested to develop tools for proving the central limit theorem in case of dependent random variables. Such techniques can be applied to our infinite-dimensional setting, which is required since the test functions form an infinite-dimensional linear space. Only the basic computational steps are explained here. A full proof would be too technical. For example, we would have to discuss the existence of solutions to (4.1). In fact, if $\beta > 1$, trajectories never touch and the solution theory is standard. But for $\beta < 1$, the crossing probability is strictly positive and the existence of solutions becomes more intricate. The proof of the limit (4.18) is based on compactness. In an appropriate function space, one ensures that along some subsequence the stochastic process of fluctuations has a limit. The limit process satisfies an equation for which uniqueness is established, which then implies convergence. We consider only the stationary process, but the time-dependent case can be handled in a similar fashion.

Using (4.2), one arrives at

$$\mathcal{L}_N \phi_N(f) = \phi_N(-V'\partial_x f + \partial_x^2 f)$$
$$+ \alpha \int_{\mathbb{R}} dx \int_{\mathbb{R}} dy \frac{f'(x) - f'(y)}{x - y} \phi_N(x) \Upsilon_N(y) = \varkappa_{[1],N}(f). \quad (4.28)$$

For the quadratic variation, we obtain

$$\mathcal{L}_N \phi_N(f)^2 - 2\phi_N(f)\mathcal{L}_N\phi_N(f) = 2N^{-1} \sum_{j=1}^{N} f'(x_j)^2 = \varkappa_{[2],N}(f). \quad (4.29)$$

As before, $\varkappa_{[1],N}(f,t)$ is the function $\varkappa_{[1],N}(f)$ evaluated at the random configuration $(x_1(t), \ldots, x_N(t))$ and the same for $\varkappa_{[2],N}(f,t)$. By the standard theory of Markov processes,

$$M_{[1],N}(f,t) = \phi_N(f,t) - \phi_N(f,0) - \int_0^t ds \varkappa_{[1],N}(f,s) \quad (4.30)$$

and

$$M_{[2],N}(f,t) = M_{[1],N}(f,t)^2 - \int_0^t ds \, \varkappa_{[2],N}(f,s) \quad (4.31)$$

are martingales.

By compactness, one first ensures that the limit in (4.18) exists. Since the martingale property is preserved in the limit $N \to \infty$, using the law of large numbers to handle $\varkappa_{[1],N}(f)$ and $\varkappa_{[2],N}(f)$, one concludes that

$$M_{[1]}(f,t) = \phi(f,t) - \phi(f,0) - \int_0^t ds\, \phi(-V'\partial_x f + \partial_x^2 f, s)$$

$$- \int_0^t ds\, \alpha \int_\mathbb{R} dx \int_\mathbb{R} dy \frac{f'(x) - f'(y)}{x - y} \phi(x,t)\varrho_\mathfrak{s}(y) \qquad (4.32)$$

and

$$M_{[2]}(f,t) = M_{[1]}(f,t)^2 - 2t \int_\mathbb{R} dx \varrho_\mathfrak{s}(x) f'(x)^2 \qquad (4.33)$$

are still martingales under the limit process $\phi(f,t)$. The unique solution to the latter martingale problem is the linear Langevin equation of (4.19), thereby confirming the claim. ◆◆

Notes and references

Section 4.0

Freeman Dyson introduced his model in Dyson (1962). An interesting account of the early history is his preface to *The Oxford Handbook on Random Matrix Theory* edited by Akemann *et al.* (2011). The handbook serves as a useful source of information and allows one to capture the vastness of the subject. An introductory monograph on random matrices is Anderson *et al.* (2010). Dyson Brownian motion also serves as a technical tool in the study of universality for Wigner random matrices, see Erdős *et al.* (2012), and the edge behavior of the density of states of the beta random matrix ensembles, see Bourgade *et al.* (2014).

Section 4.1

For general $\beta > 0$, Dyson Brownian motion is properly constructed in Cépa and Lépingle (1997), where also the law of large numbers is proved.

Section 4.2

Fluctuation theory at strong coupling is established in Israelsson (2001). The methods used can be extended to the case discussed here. Fluctuation theory for a wider class of stochastic particle systems is discussed in Spohn (1991) with more detailed arguments and references on the martingale method. For thermal equilibrium, eigenvalue fluctuations are proved by Nakano and Trinh (2018) using a moment method. Hardy and Lambert (2021) employ arguments from optimal transport to more directly estimate the distance between the true density function and the approximating Gaussian.

Chapter 5

Hydrodynamics for Hard Rods

So far our focus has been the generalized free energy. But hydrodynamics relies on further building blocks. The hard rod system will serve to illustrate the method as such, thereby building a bridge towards the Toda lattice.

5.1 Hard rod fluid

The fluid consists of hard rods of length a, $a \in \mathbb{R}$, which interact through elastic collisions, see Figure 5.1. The lattice version will be discussed in Section 5.2. Positions are constrained by $r_j \geqslant a$ and quasiparticles maintain their velocity through collisions. At a collision, i.e., $r_j = a$, q_j jumps instantaneously to $q_j + a$ and q_{j+1} to $q_{j+1} - a$. Besides the number of particles, any sum function of velocities is conserved. This field is labeled by some general function, ϕ, and has a density given by

$$Q^{[\phi]}(x) = \sum_{j \in \mathbb{Z}} \delta(q_j - x)\phi(p_j). \tag{5.1}$$

The particle number is included and corresponds to the constant function, $\phi(w) = 1$. The rod length can be negative. In fact, if $a < 0$, then for sufficiently high pressure, the free volume turns negative, a property resembling the Toda lattice.

To obtain the current, one has to differentiate the density with respect to t. Then, there is the flow term resulting from the δ-function and a collision term resulting from ϕ. For the latter, we consider a small time $dt > 0$ and the change $\phi(p_j(dt)) - \phi(p_j)$. If $r_j - a < \delta$ with sufficiently small δ and if

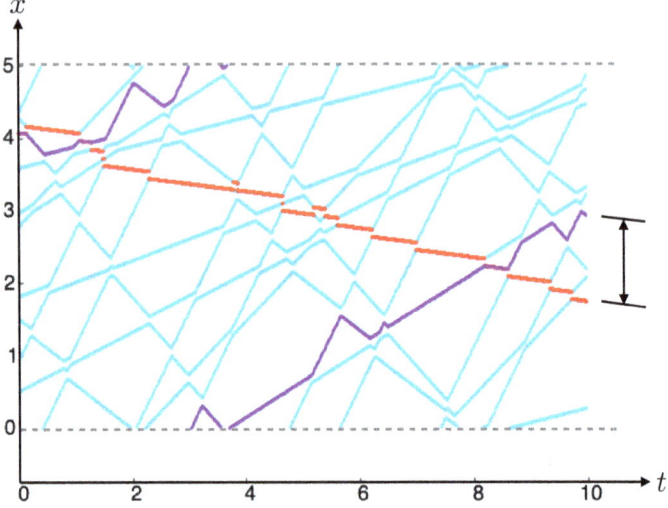

Fig. 5.1. Displayed are the trajectories, $q_j(t)$, of nine hard rod particles with length 0.3 in a volume of size 5. Very roughly the density $\bar{\rho} = 1.8$ and inverse temperature $\beta = 1$, which corresponds to a pressure $P = 3.9$. The trajectory of the eighth particle is shown in magenta and its quasiparticle trajectory in red. The scattering shift is indicated on the right. Courtesy of A. Kundu.

$p_{j+1} < p_j$, then $p_j(\mathrm{d}t) = p_{j+1}$. Similarly, if $r_{j-1} - a < \delta$ and if $p_{j-1} > p_j$, then $p_j(\mathrm{d}t) = p_{j-1}$, which implies

$$\phi(p_j(\mathrm{d}t)) - \phi(p_j) = (\delta(r_j - a)(\phi(p_{j+1}) - \phi(p_j))\chi(\{p_{j+1} < p_j\})(p_j - p_{j+1})$$
$$+ \delta(r_{j-1} - a)(\phi(p_{j-1}) - \phi(p_j))\chi(\{p_j < p_{j-1}\})(p_{j-1} - p_j)\mathrm{d}t, \qquad (5.2)$$

where $\chi(\{\cdot\})$ is the indicator function of the set $\{\cdot\}$. Thus,

$$J^{[\phi]}(x) = \sum_{j\in\mathbb{Z}}(\delta(q_j - x)p_j\phi(p_j) + \big(\theta(q_j - x) - \theta(q_{j-1} - x)\big)\delta(r_{j-1} - a)$$

$$\times \big(\phi(p_{j-1}) - \phi(p_j)\big)\chi(\{p_j < p_{j-1}\})(p_{j-1} - p_j)) \qquad (5.3)$$

with θ the step function, $\theta(x) = 0$ for $x \leqslant 0$ and $\theta(x) = 1$ for $x > 0$. Under the hard rod dynamics, the local conservation law has the standard form

$$\partial_t Q^{[\phi]}(x,t) + \partial_x J^{[\phi]}(x,t) = 0. \qquad (5.4)$$

A GGE is characterized by the one-particle distribution function $f(v) = \bar{\rho}h(v)$. Here, $h \geqslant 0$ and $\langle h \rangle = 1$. The velocities $\{p_j, j \in \mathbb{Z}\}$ are a family of i.i.d. random variables with common probability density h.

The positions $\{q_j, j \in \mathbb{Z}\}$ are statistically invariant under spatial shifts and $\{r_j = q_{j+1} - q_j\}$ are i.i.d. such that r_j is exponentially distributed with density $P \exp\left(-P(x - a)\right)\chi(\{x \geqslant a\})$, $P > 0$. Then, its first moment $\langle x \rangle = \bar{\rho} = P/(1 + aP)$. Close packing is reached in the limit $P \to \infty$ with $\bar{\rho} = 1/a$. In the sense of thermodynamics, $\bar{\rho}, h$ are extensive parameters, while P, V are intensive parameters upon setting $h = \mathrm{e}^{-V}/\langle \mathrm{e}^{-V} \rangle$. Slightly deviating from our standard convention the GGE average is denoted by $\langle \cdot \rangle_{\bar{\rho},h}$. The required GGE averages are easily computed. For the fields, one obtains

$$\langle Q^{[\phi]}(0) \rangle_{\bar{\rho},h} = \bar{\rho}\langle h\phi \rangle. \tag{5.5}$$

For the currents, one uses that

$$\int_{\mathbb{R}} \mathrm{d}v \int_{\mathbb{R}} \mathrm{d}w h(v) h(w) \big(\phi(w) - \phi(v)\big) \chi(\{v < w\})(w - v)$$

$$= \int_{\mathbb{R}} \mathrm{d}v \phi(v)(v - u) h(v) \tag{5.6}$$

with the mean velocity $u = \langle vh(v) \rangle$. Then,

$$\langle J^{[\phi]}(0) \rangle_{\bar{\rho},h} = \bar{\rho}\langle hv^{\mathrm{eff}}\phi \rangle \tag{5.7}$$

with the effective velocity

$$v^{\mathrm{eff}}(v) = \frac{1}{1 - a\bar{\rho}}(v - a\bar{\rho}u). \tag{5.8}$$

The hydrodynamic evolution equations are built from the average conserved fields and currents, i.e., the inputs (5.5) and (5.7). For the initial state, one assumes to have locally the statistics of some GGE. In our case, one example would be to maintain independence but let $f = \bar{\rho}h$ change very slowly on the scale set by the maximal correlation length. On a more formal level, one introduces the dimensionless parameter ϵ, $\epsilon \ll 1$, of a typical scale for the ratio "microscopic length/macroscopic length". Microscopically, the physically relevant length is the maximal correlation length of the initial GGE. Translated to the initial f, the scale separation is ensured through the functional form $f(\epsilon x; v)$. Given such random initial conditions, the hard rod fluid is evolving through free motion and collisions. Independence of momenta is lost immediately. Yet on the macroscopic scale, nothing is moving, at least not for short times. Due to the conservation laws, changes can be observed only on microscopic times of order $\epsilon^{-1}t$. Hydrodynamic scaling is ballistic since changes in space and in time are both $\mathcal{O}(\epsilon^{-1})$ in microscopic units. GHD then postulates that even for such long times local

GGE remains approximately valid. If propagation of local GGE holds, one can write down the equations governing the evolution of the parameters characterizing local GGEs.

The starting point is the exact microscopic conservation law (5.4). Since this law holds for any dynamical trajectory, one can average over an arbitrary ensemble, in particular, the initial slowly varying state. Instead of averaging the time t fields over initial conditions, we average the physical fields over the statistical state at time t, average denoted by $\langle \cdot \rangle_{t,\epsilon}$ with the subscript ϵ indicating the parameter of initial slow variation,

$$\partial_t \langle Q^{[\phi]}(x) \rangle_{t,\epsilon} + \partial_x \langle J^{[\phi]}(x) \rangle_{t,\epsilon} = 0, \tag{5.9}$$

which is still exact. Now, switching to macroscopic spacetime $(\epsilon^{-1}x, \epsilon^{-1}t)$ propagation of local GGE amounts to the approximation

$$\langle Q^{[\phi]}(\epsilon^{-1}x) \rangle_{\epsilon^{-1}t,\epsilon} \simeq \langle Q^{[\phi]}(0) \rangle_{f(x,t)},$$
$$\langle J^{[\phi]}(\epsilon^{-1}x) \rangle_{\epsilon^{-1}t,\epsilon} \simeq \langle J^{[\phi]}(0) \rangle_{f(x,t)}, \tag{5.10}$$

where on the right-hand side $Q^{[\phi]}(x)$ is replaced by $Q^{[\phi]}(0)$, resp. $J^{[\phi]}(x)$ by $J^{[\phi]}(0)$, because of the spatial translation invariance of the GGE average $\langle \cdot \rangle_{f(x,t)}$. Using the effective velocity (5.7) evaluated at (x,t), we arrive at a closed equation for $f(t) = \bar{\rho}(t)h(t)$,

$$\partial_t f(x,t;v) + \partial_x \big((1 - a\bar{\rho}(x,t))^{-1}(v - a\bar{\rho}(x,t)u(x,t))f(x,t;v) \big) = 0. \tag{5.11}$$

Here, the arguments of f indicate that (x,t) is the spacetime patch under consideration and v is the label of the conserved field in that patch. Equation (5.11) is the hydrodynamic equation of the hard rod fluid. Through nonlinearity, the hydrodynamic fields are coupled and decouple only in the ideal gas limit $a \to 0$. For $a \geqslant 0$, the function $f(t)$ counts the number of particles in the macroscopic volume element $dxdv$ at time t. But for $a < 0$, the density $\bar{\rho}(t)$ can have negative values and so does $f(t)$.

On purpose we used the symbol f to emphasize that this quantity is very close to the familiar Boltzmann f-function in the sense of counting the number of particles in one-particle phase space. In kinetic theory and in GHD, the f-function is approximately governed by an autonomous evolution equation which is local in (x,t). In kinetic theory, the collision operator generates entropy, the motion is thus dissipative, while GHD is a coupled system of local conservation laws for which entropy is conserved. The mathematical structure of the respective dynamical evolution equations is fairly distinct.

As an apparently general feature of generalized hydrodynamics, the hydrodynamic equation (5.11) can be written in a quasilinear form. For hard rods, the transformation reads

$$\rho_{\mathsf{n}}(v) = \frac{1}{\nu - a}f(v), \quad \nu\bar{\rho} = 1, \tag{5.12}$$

and

$$\partial_t \rho_{\mathsf{n}}(x, t; v) + v^{\mathrm{eff}}(x, t; v)\partial_x \rho_{\mathsf{n}}(x, t; v) = 0. \tag{5.13}$$

The fields $\rho_{\mathsf{n}}(v)$ are the normal modes of the hard rod fluid equations and the convective equation (5.13) identifies their local propagation velocities as $v^{\mathrm{eff}}(v)$, which depend nonlinearly on ρ_{n} through (5.8).

Hydrodynamic equations are expected to become exact in the limit $\epsilon \to 0$. But in practice, hydrodynamics becomes applicable much before the actual limit. The precise value of ϵ depends on the particular physical system and its initial conditions. The range of validity for hydrodynamics is mostly a qualitative rule of thumb, rather than a sharp error estimate.

♦♦ *Hydrodynamics of simple fluids, the issue of mathematical rigor*: In 1757, Leonhard Euler published his monumental treatise on fluids, which in particular included the compressible Euler equations. In this case, the locally conserved fields are number, momentum, and total energy. Euler argued for the appropriate form of the currents on a phenomenological basis. It was a triumph of early Statistical Mechanics to understand that these currents can also be obtained from averaging the microscopic currents over a Gibbs ensemble. The required technical tool is an identity known as virial theorem. Since the resulting equations agreed with what had been known already for a long time, the discussion was mostly confined to the textbook level. Still, the derivation was a further compelling evidence for the universal validity of Statistical Mechanics. For integrable many-body systems, one cannot rely on phenomenological evidence. The only available method is to figure out the GGE averaged currents as based on the microscopic model.

A much harder question is the propagation of local GGEs. For the hard rod fluid, Roland Dobrushin and collaborators proved such a result in the mid-1980s. But for more complicated integrable systems, in particular the Toda lattice, no results are available yet. Due to their physical importance, for simple fluids, there have been many attempts to establish the compressible Euler equations in a mathematically convincing way. From that perspective, the best result is still the work of Olla, Varadhan, and Yau. They investigated the kind of dynamical mixing which would be needed for propagation of local equilibrium. In fact, they modified the mechanical model by assuming that the deterministic collisions between particles are substituted by random ones, still respecting the mechanical conservation laws. Thus, the macroscopic equations are the compressible version of Euler equations. For such stochastic dynamics, Olla, Varadhan, and Yau prove the required mixing. But for a mechanical system, mixing would have to come from deterministic chaos. The resulting gap is so huge that progress is unlikely, even allowing for a long-term perspective. ◆◆

5.2 Hard rod lattice

Physically, the fluid picture can be more easily visualized. But as in the case of the Toda chain, one can switch to the lattice version. Then, the dynamics is formulated in terms of the stretches r_j, where $r_j(t) \geqslant a$ because of the hard core. If $p_j(t) - p_{j+1}(t) < 0$, then $r_j(t)$ is decreasing and reaches collision at $r_j(t_c) = a$, t_c the collision time. Instantaneously, the momenta (p_j, p_{j+1}) are exchanged and $r_j(t)$ is increasing for $t > t_c$. Such local collisions happen for every spatial patch throughout time. Obviously, the densities of the conserved fields are

$$Q_j^{[0]} = r_j, \quad Q_j^{[\phi]} = \phi(p_j). \tag{5.14}$$

Note that $\phi = 1$ is just the constant function, hence not admissible. Thus, in contrast to the fluid, the index 0 will have to be treated differently from $n \geqslant 1$. The currents are

$$J_j^{[0]} = -p_j \tag{5.15}$$

and

$$J_j^{[\phi]} = \delta(r_{j-1} - a)(\phi(p_{j-1}) - \phi(p_j))\chi(\{p_j < p_{j-1}\})(p_{j-1} - p_j). \tag{5.16}$$

Compared to (5.3), there is no flow term and the factor $(\theta(q_j - x) - \theta(q_{j-1} - x))$ is missing.

For a hard rod lattice GGE, the velocities are i.i.d. with probability density $h(v)$, $h \geqslant 0$, $\langle h \rangle = 1$ and the r_js are also i.i.d. with probability density $P \exp(-P(x - a))\chi(\{x \geqslant a\})$, $P > 0$. The average stretch is then $\nu = P^{-1} + a$. In particular, $\nu - a > 0$. Now, ν, h, are the extensive parameters, while P, V are still intensive parameters. The GGE average is denoted by $\langle \cdot \rangle_{\nu,h}$. The averaged conserved fields are given by

$$\langle Q_0^{[0]} \rangle_{\nu,h} = \nu, \quad \langle Q_0^{[\phi]} \rangle_{\nu,h} = \langle h\phi \rangle, \tag{5.17}$$

and the averaged currents by

$$\langle J_0^{[0]}\rangle_{\nu,h} = -\langle hv\rangle = -u,$$

$$\langle J_0^{[\phi]}\rangle_{\nu,h} = (\nu - a)^{-1}\left(\langle hv\phi\rangle - u\langle h\phi\rangle\right) = \nu^{-1}\langle h(v^{\mathrm{eff}} - u)\phi\rangle, \qquad (5.18)$$

where we used the same computation as in (5.6) and in Eq. (5.8) rewritten as $v^{\mathrm{eff}}(v) = (\nu - a)^{-1}(\nu v - au)$. As a result, the hydrodynamic equations of the hard rod lattice take the form

$$\partial_t \nu(x,t) - \partial_x u(x,t) = 0,$$

$$\partial_t h(x,t;v) + \partial_x\left(\nu(x,t)^{-1}(v^{\mathrm{eff}}(x,t;v) - u(x,t))h(x,t;v)\right) = 0. \qquad (5.19)$$

Compared to (5.11), there are now two equations rather than one. But switching to normal modes through

$$\rho_{\mathsf{n}}(v) = \frac{1}{\nu - a}h, \qquad (5.20)$$

one arrives again at the quasilinear equation (5.13).

5.3 TBA, collision rate ansatz

A central model-dependent feature of TBA is the two-particle scattering shift as appearing in the definition of the T operator. Now for hard rods, the two-particle scattering shift equals $-a$. To have the same structure as the Toda lattice would mean to replace T in (3.50) by the projection operator

$$T = -a|1\rangle\langle 1|, \qquad (5.21)$$

in other words, $Tf(w) = -a\langle f\rangle$. With this input, the TBA equation for the hard rod lattice reads

$$V - \mu + a\langle\rho_{\mathsf{n}}\rangle + \log\rho_{\mathsf{n}} = 0, \qquad \langle\rho_{\mathsf{n}}\rangle = P. \qquad (5.22)$$

Since T is such a simple operator, the solution is still explicit with

$$\mu = \log P + aP - \log\langle e^{-V}\rangle \qquad (5.23)$$

and

$$h = e^{-V}/\langle e^{-V}\rangle, \qquad \rho_{\mathsf{n}} = (\nu - a)^{-1}h, \qquad \rho_{\mathsf{p}} = \nu^{-1}h. \qquad (5.24)$$

The average of the conserved fields is determined by

$$\langle Q_0^{[\phi]}\rangle_{\nu,h} = \langle h\phi\rangle = \nu\langle\rho_{\mathsf{p}}\phi\rangle, \qquad (5.25)$$

in agreement with (5.17). Indeed, with adjusted T, the TBA formalism holds also for hard rods.

Qualitatively, the Toda lattice is best approximated by choosing $a < 0$. The identities in (5.25) then provide a simple illustration for the uniqueness Insert below Eq. (3.62). The stretch $\nu(P) = (1/P) + a$ is monotone decreasing with $P_c = -1/a$. Since $\nu - a > 0$, in addition $\rho_n \geqslant 0$. However, ρ_p globally flips sign at P_c, while $\nu\rho_p \geqslant 0$.

We now invert the logic and try to guess a formula for the GGE averaged currents of hard rods, through using a conventional GHD argument. Employed is the notion of quasiparticles, most easily visualized in the fluid picture. A quasiparticle retains its velocity when undergoing a collision. While a physical particle rattles back and forth between its two neighbors, the quasiparticle moves with constant velocity interrupted by jumps of size $\pm a$ at collisions. We now prepare a GGE characterized by $\bar{\rho}, h$ and initially start a tracer quasiparticle at the origin with velocity v. Viewed on a somewhat coarser scale, when time-averaging over many collisions, the quasiparticle travels with a yet to be determined effective velocity $v^{\text{eff}}(v)$, which turns out to be increasing in v. The tracer quasiparticle collides with a fluid quasiparticle of velocity w. Such fluid particles have density $\rho_p(w)$. If the collision is from the left, $v < w$, then the tracer jumps by $-a$. Hence, the collision rate is $\rho_p(w)(v^{\text{eff}}(w) - v^{\text{eff}}(v))$. On the other hand, if the collision is from the right, $v > w$, then the tracer jumps by a and the collision rate is $\rho_p(w)(v^{\text{eff}}(v) - v^{\text{eff}}(w))$. When integrating over all fluid quasiparticles, the bare tracer quasiparticle velocity is modified through collisions according to

$$v^{\text{eff}}(v) = v - a \int_v^\infty \mathrm{d}w \rho_p(w)\big(v^{\text{eff}}(w) - v^{\text{eff}}(v)\big)$$

$$+ a \int_{-\infty}^v \mathrm{d}w \rho_p(w)\big(v^{\text{eff}}(v) - v^{\text{eff}}(w)\big) \tag{5.26}$$

and thus

$$v^{\text{eff}}(v) = v - a \int_{\mathbb{R}} \mathrm{d}w \rho_p(w)\big(v^{\text{eff}}(w) - v^{\text{eff}}(v)\big). \tag{5.27}$$

Equation (5.27) is the *collision rate ansatz*. More concisely in operator form,

$$v^{\text{eff}}(v) = v + T(\rho_p v^{\text{eff}})(v) - (T\rho_p)(v)v^{\text{eff}}(v). \tag{5.28}$$

For hard rods, the rate equation is easily solved with the result

$$v^{\text{eff}}(v) = v + \frac{a\bar{\rho}}{1 - a\bar{\rho}} \int_{\mathbb{R}} \mathrm{d}w h(w)(v - w) = (1 - a\bar{\rho})^{-1}(v - a\bar{\rho}u). \tag{5.29}$$

Indeed, the average current equals $\bar{\rho}hv^{\text{eff}}$, in agreement with the microscopic computation (5.8).

With hindsight, the expression (5.27) for the effective velocity carries already a recipe of how to extend to other integrable models. The two-particle scattering shift for the Toda fluid equals $2 \log |v - v'|$. Thus, one might hope to obtain the correct average currents for the Toda fluid by considering the motion of a tracer quasiparticle. Analytically, this amounts to substituting the operator T in (5.28) by the operator T from (3.55).

Notes and references

Section 5.1

The hydrodynamic limit for hard rods, i.e., the asymptotic validity of (5.11), is proved by Boldrighini *et al.* (1983) under the assumption of a sufficiently random initial state. The limit holds with probability one. A short account is provided by Spohn (1991). A different proof is constructed by Ferrari *et al.* (2023). From the perspective of generalized hydrodynamics, hard rods are discussed by Doyon and Spohn (2017a) and Doyon (2020). In experiments on integrable many-body systems, particles are moving in a one-dimensional tube and are confined by an external potential along the tube. Generically, this breaks integrability. To elucidate more precisely the underlying mechanism, hard rods serve as an instructive model, see the contribution by Cao *et al.* (2018).

For Hamiltonian many-particle systems with short-range interactions, Morrey (1955) launched an interesting early attempt. The modification of adding randomness in collisions is studied by Olla *et al.* (1993). The hydrodynamic limit on the Euler scale can be viewed as a stability result, in the sense that on the ballistic scale there is no deviation yet from local equilibrium, resp. local GGE.

Starting in the early 1980s, it was realized that the hydrodynamic limit is also a challenging problem for stochastic lattice gases and interacting diffusions. The dynamics is overdamped in the sense that particles have only a position and evolve through random update rules. An early account is Spohn (1991).

Chapter 6

Equations of Generalized Hydrodynamics

6.1 Average currents

The microscopic version of the currents has already been obtained, see (2.25) and (2.27). Now, the goal is to compute their GGE average in the limit $N \to \infty$. As before, the lattice size is N and we use the pressure ensemble (3.6). Then, in terms of the spectral resolution of L_N, the nth microscopic current reads

$$J^{[n],N} = \mathrm{tr}\big[(L_N)^n L_N^{\downarrow}\big] = N \langle \rho_{\mathrm{J},N} \varsigma_n \rangle, \tag{6.1}$$

$n \geqslant 1$, with

$$\rho_{\mathrm{J},N}(w) = \frac{1}{N} \sum_{j=1}^{N} \delta(w - \lambda_j) \left(\sum_{i=1}^{N} a_i \psi_j(i) \psi_j(i+1) \right). \tag{6.2}$$

The δ-peaks of the Lax density of states are weighted by coefficients with arbitrary sign. In analogy, not to duplicate names, we call $\rho_{\mathrm{J},N}$ the *empirical current DOS*. For given GGE, the current DOS is self-averaging and has the deterministic limit ρ_{J}. In particular,

$$\lim_{N \to \infty} \frac{1}{N} \langle J^{[n],N} \rangle_{P,V,N} = \langle J_0^{[n]} \rangle_{P,V} = \langle \rho_{\mathrm{J}} \varsigma_n \rangle. \tag{6.3}$$

Setting $n = 0$ on the right-hand side of Eq. (6.1), one concludes $\langle \rho_{\mathrm{J}} \rangle = 0$. Hence, as physically expected, ρ_{J} cannot have a definite sign. From the few numerical simulations available, very qualitatively, $\rho_{\mathrm{J}}(w)$ has the shape of $-\partial_w \rho_{\mathrm{Q}}(w)$.

Since the Dumitriu–Edelman identity worked so well for the conserved fields, one is tempted to use the same strategy for the currents. But now not

75

only the distribution of eigenvalues is in demand. While we have an explicit form for the joint distribution of $\{\lambda_j, \psi_j(1), j = 1, \ldots, N\}$, to apply the map Φ of Dumitriu–Edelman to the weights $\sum_{i=1}^{N} a_i \psi_j(i) \psi_j(i+1)$ seems to be a complicated enterprise. A new approach is needed. Unexpectedly, the key idea will come from the susceptibility matrices.

We start with the fields and define the infinite volume field–field correlator

$$C_{m,n}(j - i) = \langle Q_j^{[m]} Q_i^{[n]} \rangle_{P,V}^{c} \tag{6.4}$$

for $m, n \geqslant 0$, where the superscript c denotes truncation, respectively, connected correlation, $\langle gf \rangle_{P,V}^{c} = \langle gf \rangle_{P,V} - \langle g \rangle_{P,V} \langle f \rangle_{P,V}$. Truncated correlations decay rapidly to zero and the field–field susceptibility matrix is given by

$$C_{m,n} = \sum_{j \in \mathbb{Z}} C_{m,n}(j) = \langle Q^{[m]}; Q^{[n]} \rangle_{P,V}, \tag{6.5}$$

where the second equality is merely a convenient notation. $C_{m,n}$ is the matrix of second derivatives of the generalized free energy. Correspondingly, we introduce the field–current correlator and the field–current susceptibility matrix

$$B_{m,n}(j - i) = \langle J_j^{[m]} Q_i^{[n]} \rangle_{P,V}^{c}, \quad B_{m,n} = \sum_{j \in \mathbb{Z}} B_{m,n}(j). \tag{6.6}$$

Despite its apparently asymmetric definition, B satisfies

$$B_{m,n}(j) = B_{n,m}(-j). \tag{6.7}$$

To prove, we employ the conservation law and spacetime stationarity to arrive at

$$\partial_j \langle J_j^{[m]}(t) Q_0^{[n]}(0) \rangle_{P,V}^{c} = -\partial_t \langle Q_j^{[m]}(t) Q_0^{[n]}(0) \rangle_{P,V}^{c}$$

$$= -\partial_t \langle Q_0^{[m]}(0) Q_{-j}^{[n]}(-t) \rangle_{P,V}^{c} = \partial_j \langle Q_0^{[m]}(0) J_{-j}^{[n]}(-t) \rangle_{P,V}^{c}, \tag{6.8}$$

denoting the difference operator by $\partial_j f(j) = f(j+1) - f(j)$. Setting $t = 0$, the difference $\langle J_j^{[m]} Q_0^{[n]} \rangle_{P,V}^{c} - \langle J_{-j}^{[m]} Q_0^{[m]} \rangle_{P,V}^{c}$ is constant in j. Since truncated correlations decay to zero, this constant has to vanish, which yields (6.7). In particular, the field–current susceptibility matrix is symmetric:

$$B_{m,n} = B_{n,m}. \tag{6.9}$$

Using this symmetry and restricting to $n \geqslant 1$, we consider the P-derivative of the average current

$$\partial_P \langle (L^n L^{\downarrow})_{0,0} \rangle_{P,V} = -B_{n,0} = -B_{0,n} = \langle Q^{[1]}; Q^{[n]} \rangle_{P,V}, \tag{6.10}$$

since $J^{[0]} = -Q^{[1]}$ by (2.27). We easily arrived at a very surprising identity. The P-derivative of the average current equals a particular matrix element of the field–field susceptibility. Such susceptibility depends only on the eigenvalues of the Lax matrix. In fact, this quantity has been studied already in Section 4.1 and one only has to adjust the results from there, setting $\alpha = P$.

As in the case of the free energy, since the pressure is varying as $1/N$, the fluctuation covariance is adding up, resulting in

$$\langle \varsigma_1, C^\sharp \varsigma_n \rangle = \int_0^1 du \langle Q^{[1]}; Q^{[n]} \rangle_{uP,V}, \tag{6.11}$$

with the operator $C^\sharp$ defined in (4.23). Inserting from (6.10), one concludes that

$$\partial_P \left(\langle J_0^{[n]} \rangle_{P,V} - P \langle \varsigma_1, C^\sharp \varsigma_n \rangle \right) = 0. \tag{6.12}$$

The term in the round brackets has to be independent of P, in particular equal to its value at $P = 0$. Since the covariance is bounded, the second summand vanishes at $P = 0$. For $\langle J_0^{[n]} \rangle_{P,V}$, we use the random walk expansion (2.18). Each walk contains at least one factor of a_j and thus vanishes because under the *a priori* measure $a_j \to 0$ in the limit $P \to 0$. Therefore,

$$\langle J_0^{[n]} \rangle_{P,V} = P \langle \varsigma_1, C^\sharp \varsigma_n \rangle. \tag{6.13}$$

Inserting from (4.23) and substituting as $\alpha \rho_\mathsf{s} = \rho_\mathsf{n}$,

$$\langle J_0^{[n]} \rangle_{P,V} = \langle \varsigma_1, (1 - \rho_\mathsf{n} T)^{-1} \rho_\mathsf{n} \varsigma_n \rangle$$
$$- \nu \langle \varsigma_1 (1 - \rho_\mathsf{n} T)^{-1} \rho_\mathsf{n} \rangle \langle \varsigma_n (1 - \rho_\mathsf{n} T)^{-1} \rho_\mathsf{n} \rangle. \tag{6.14}$$

Since $(1 - \rho_\mathsf{n} T)^{-1} \rho_\mathsf{n}$ is a symmetric operator, one arrives at

$$\langle J_0^{[n]} \rangle_{P,V} = \langle \rho_\mathsf{n} \varsigma_1^{\mathrm{dr}} \varsigma_n \rangle - q_1 \langle \rho_\mathsf{p} \varsigma_n \rangle, \quad q_1 = \nu \langle \rho_\mathsf{p} \varsigma_1 \rangle. \tag{6.15}$$

Using (6.3), one finally concludes

$$\langle \rho_\mathsf{J} \varsigma_n \rangle = \langle (\rho_\mathsf{n} \varsigma_1^{\mathrm{dr}} - q_1 \rho_\mathsf{p}) \varsigma_n \rangle \tag{6.16}$$

valid for $n \geqslant 1$ and, in addition,

$$\langle \rho_\mathsf{J} \varsigma_0 \rangle = -q_1. \tag{6.17}$$

In the previous chapter, we entertained the possibility that, by analogy, the Toda current can be written as

$$\rho_\mathsf{J} = \rho_\mathsf{p}(v^{\mathrm{eff}} - q_1), \tag{6.18}$$

compare with (5.18) and (5.24). Using the scattering shift of the Toda lattice, this proposal amounts to

$$v^{\text{eff}}(v) = v + 2 \int_{\mathbb{R}} dw \log |v - w| \rho_{\mathsf{p}}(w) \big(v^{\text{eff}}(w) - v^{\text{eff}}(v) \big). \tag{6.19}$$

If such an analogy holds, then $\rho_{\mathsf{p}} v^{\text{eff}} = \rho_{\mathsf{n}} \varsigma_1^{\text{dr}}$, and hence,

$$v^{\text{eff}} = \frac{\varsigma_1^{\text{dr}}}{\varsigma_0^{\text{dr}}}. \tag{6.20}$$

To confirm, we start from

$$
\begin{aligned}
T(\rho_{\mathsf{p}} v^{\text{eff}}) - (T \rho_{\mathsf{p}}) v^{\text{eff}} &= T(\rho_{\mathsf{n}} \varsigma_1^{\text{dr}}) - (T \rho_{\mathsf{p}}) v^{\text{eff}} \\
&= (T \rho_{\mathsf{n}} - 1 + 1)(1 - T \rho_{\mathsf{n}})^{-1} \varsigma_1 - (T \rho_{\mathsf{p}}) v^{\text{eff}} \\
&= -\varsigma_1 + \varsigma_1^{\text{dr}} - \big(\rho_{\mathsf{n}}^{-1} \rho_{\mathsf{p}} - 1 \big) v^{\text{eff}} = v^{\text{eff}} - \varsigma_1, \tag{6.21}
\end{aligned}
$$

as claimed.

At low pressure, the Toda particles are far apart and interact mostly through isolated two-particle collisions. Quasiparticles are then defined as for hard rods, i.e., by maintaining their velocity upon excluding the time span for collisions. The tracer quasiparticle jumps by a distance regulated by the Toda two-particle scattering shift. The validity of the collision rate ansatz (6.19) can be argued as we did already for a hard rod fluid. However, for a dense Toda fluid with P not too close to 0, the notion of a quasiparticle is fuzzy, even more so the meaning of a tracer quasiparticle. The validity of (6.19) for arbitrary GGE parameters is concluded only indirectly.

Conceptually, the collision rate ansatz can be read in a somewhat different way, as inspired by the time-of-flight method employed in recent experiments on the rapidity distribution for the δ-Bose gas. One prepares the N-particle Toda system in a GGE and runs the dynamics over the time span $[0, t]$ with periodic boundary conditions, t, of the same order as N. To identify the tracer quasiparticle, a past and future diagnostic step is added, to say in the time interval $[-\tau, 0]$ backward and $[t, t + \tau]$ forward, the dynamics is executed with open boundary conditions, $\tau > 0$, and of order 1. Then, at times $-\tau$, $t + \tau$, particles are well separated and have definite velocities. At time $-\tau$, we pick one particle with velocity w and call it tracer quasiparticle. Then, at time $t + \tau$, there is exactly one particle with the same velocity w. For $\tau \ll t$, the displacement of the tracer quasiparticle should be $v^{\text{eff}}(w)t$ in good approximation.

6.2 Hydrodynamic equations

On the hydrodynamic scale, the local GGE is characterized by the stretch ν and the Lax DOS $\nu\rho_{\mathsf{p}}$, both of which now become spacetime-dependent. Merely inserting the average currents, one arrives at the Euler-type hydrodynamic evolution equations,

$$\partial_t \nu(x,t) - \partial_x q_1(x,t) = 0, \tag{6.22}$$

$$\partial_t \langle \nu(x,t)\rho_{\mathsf{p}}(x,t)\varsigma_n \rangle + \partial_x \langle (v^{\text{eff}}(x,t) - q_1(x,t))\rho_{\mathsf{p}}(x,t)\varsigma_n \rangle = 0,$$

for $n \geqslant 1$. The latter equation extends to $n = 0$, since the bracket under ∂_t equals 1 and, as can be concluded from (6.21), that under ∂_x equals 0. Therefore, one can switch to the pointwise version of the hydrodynamic equations,

$$\partial_t \nu(x,t) - \partial_x q_1(x,t) = 0, \tag{6.23}$$

$$\partial_t \big(\nu(x,t)\rho_{\mathsf{p}}(x,t;v) \big) + \partial_x \big((v^{\text{eff}}(x,t;v) - q_1(x,t))\rho_{\mathsf{p}}(x,t;v) \big) = 0.$$

As a most remarkable feature of generalized hydrodynamics, these equations can be transformed explicitly to a quasilinear form. For this purpose, we recall the identity (3.61), which expresses ρ_{n} in terms of ρ_{p},

$$\rho_{\mathsf{n}} = \frac{\rho_{\mathsf{p}}}{1 + (T\rho_{\mathsf{p}})}. \tag{6.24}$$

Then, Eq. (6.23) acquires the normal form

$$\partial_t \rho_{\mathsf{n}} + \nu^{-1}(v^{\text{eff}} - q_1)\partial_x \rho_{\mathsf{n}} = 0. \tag{6.25}$$

Thus, the linearization operator is in fact merely a multiplication by $\nu^{-1}(v^{\text{eff}} - q_1)$, in other words, the operator is diagonal. For nonintegrable chains, the corresponding operator is a 3×3 matrix, which in general is not diagonal. Its diagonalization would yield the normal modes. The solution of the respective hyperbolic conservation laws can develop shock discontinuities, since locally the solution may have to jump to another eigenvalue. In our case, $\nu^{-1}(v^{\text{eff}} - q_1)$ is expected to be a smooth function. Thus, the spectrum of the linearization operator is continuous and has no eigenvalues. Hence, for smooth initial data, solutions of (6.25) are expected to stay smooth. No mathematical analysis has been attempted to verify this conjecture. In fact, the situation could be more subtle since ν can take either sign and ν^{-1} might be singular.

To verify (6.25), we start from

$$\rho_{\mathsf{p}}\partial_t \nu + \nu\partial_t\rho_{\mathsf{p}} + \partial_x\big((v^{\mathrm{eff}} - q_1)\rho_{\mathsf{p}}\big) = 0, \tag{6.26}$$

which together with the continuity equation yields

$$\nu\partial_t\rho_{\mathsf{p}} + \partial_x(v^{\mathrm{eff}}\rho_{\mathsf{p}}) - q_1\partial_x\rho_{\mathsf{p}} = 0. \tag{6.27}$$

Using this identity and (6.24), one obtains

$$\nu\partial_t\rho_{\mathsf{n}} = \frac{\rho_{\mathsf{n}}}{\rho_{\mathsf{p}}}\big(-\partial_x(v^{\mathrm{eff}}\rho_{\mathsf{p}}) + q_1\partial_x\rho_{\mathsf{p}}\big) - \frac{\rho_{\mathsf{n}}^2}{\rho_{\mathsf{p}}}\big(T(-\partial_x(v^{\mathrm{eff}}\rho_{\mathsf{p}}) + q_1\partial_x\rho_{\mathsf{p}})\big)$$

$$= \frac{\rho_{\mathsf{n}}}{\rho_{\mathsf{p}}}\big(-\partial_x(v^{\mathrm{eff}}\rho_{\mathsf{p}}) + \rho_{\mathsf{n}}\partial_x T(v^{\mathrm{eff}}\rho_{\mathsf{p}})\big) + q_1\partial_x\rho_{\mathsf{n}}. \tag{6.28}$$

The effective velocity solves the integral equation

$$v^{\mathrm{eff}} = v + T(v^{\mathrm{eff}}\rho_{\mathsf{p}}) - (T\rho_{\mathsf{p}})v^{\mathrm{eff}} \tag{6.29}$$

and, by inserting (6.24),

$$\partial_x\big(\rho_{\mathsf{n}}^{-1}v^{\mathrm{eff}}\rho_{\mathsf{p}}\big) = \partial_x T(v^{\mathrm{eff}}\rho_{\mathsf{p}}). \tag{6.30}$$

Hence, the first summand in (6.28) reads

$$\frac{\rho_{\mathsf{n}}}{\rho_{\mathsf{p}}}\big(-\partial_x(v^{\mathrm{eff}}\rho_{\mathsf{p}}) + \rho_{\mathsf{n}}\partial_x(\rho_{\mathsf{n}}^{-1}v^{\mathrm{eff}}\rho_{\mathsf{p}})\big) = -v^{\mathrm{eff}}\partial_x\rho_{\mathsf{n}}, \tag{6.31}$$

thereby confirming (6.25).

The normal form (6.25) of the hydrodynamic equations is a crucial insight. *A priori*, the coupled conservation laws might be so complicated that it becomes an impossible task to extract information of interest. Here, the normal form is of considerable help. It allows us to obtain some partially analytic solutions, as the domain wall problem and the linearized version to be discussed in the following chapters. The normal form also suggests how to devise a numerical scheme for solving the hydrodynamic equations. In its most basic version, at the current time, t, one keeps $\nu^{-1}(v^{\mathrm{eff}} - q_1)$ fixed and solves the resulting linear equation for one further time step dt. With the so obtained $\rho_{\mathsf{n}}(t + dt)$, one updates $\nu^{-1}(v^{\mathrm{eff}} - q_1)$ according to (6.20) for v^{eff}, $\nu^{-1} = \langle\rho_{\mathsf{p}}\rangle$, $\nu^{-1}q_1 = \langle\rho_{\mathsf{p}}\varsigma_1\rangle$, and $\rho_{\mathsf{p}} = \rho_{\mathsf{n}}\varsigma_0^{\mathrm{dr}}$. No explicit use of TBA is involved in this step. The iteration is then repeated many times until the desired final time is reached.

The collision rate equation suggests a distinct approach called the flea gas algorithm. The basic idea is to replace the true motion by an instantaneous exchange of velocities v_1, v_2 whenever the two particles first reach the distance $2 \log |v_1 - v_2|$. Thereby, the simulation of the particle dynamics is massively speeded up and it is ensured that on large scales, the Toda generalized hydrodynamics is realized. In fact, the true dynamics is more complicated, since a third particle or even more particles might interfere with the collision between particles 1 and 2. Properly setting up the particle model requires a separation into isolated particle clusters combined with an instantaneous many-body collision. For the Toda fluid, and also integrable quantum many-body systems, one can use such a scheme as an alternative to direct numerical simulations of the hydrodynamic equations.

Physically, the initial conditions have to satisfy the constraints resulting from GGE. This amounts to $\nu \in \mathbb{R}$, $\nu \rho_p \geqslant 0$, and $\rho_n \geqslant 0$. Such conditions should be propagated by the Euler equations. For ρ_n, this property follows from (6.25). The prefactor of ∂_x looks singular at $\nu = 0$ and propagating properties of ρ_p are less obvious. A specific example is discussed in Chapter 8.

Comparing Eqs. (5.19) and (6.23) and identifying $h = \nu \rho_p$, $u = q_1$, Toda and hard rods have structurally identical hydrodynamic evolution equations. On the hard rod side, the hydrodynamic limit has been proved, thereby supporting the validity of the corresponding property for the Toda lattice.

Notes and references

Section 6.1

The computation leading to (6.8) is taken from De Nardis *et al.* (2019), see also Karevski and Schütz (2019). The symmetry of the B-matrix is discussed in Grisi and Schütz (2011) and Spohn (2014). It is rather likely that there is earlier work. The connection between the symmetry of the B-matrix and the average current was first noted by Spohn (2020a) and extended to quantum systems by Yoshimura and Spohn (2020). For integrable quantum systems, the appropriately adjusted rate equation (6.19) has been discovered by Castro-Alvaredo *et al.* (2016). Very quickly, it was understood that the scheme originally developed in Bertini *et al.* (2016) and Castro-Alvaredo *et al.* (2016) for specific models has a much wider applicability, see Doyon *et al.* (2018). There are further attempts to microscopically justify the form of the average currents: to mention a few, see the ones by Vu and Yoshimura (2019), Borsi *et al.* (2021), Pristyák and Pozsgay (2023), and the review by Cubero *et al.* (2021).

Section 6.2

The transformation to normal form is established by Castro-Alvaredo *et al.* (2016). The flea gas algorithm is discussed by Doyon *et al.* (2018) and Mestyán and Alba (2020) as an alternative approach to more standard PDE discretization schemes. A numerically oriented contribution is Møller *et al.* (2020). In Bulchandani (2017), it is argued that GHD equations are continuum integrable systems, which is related to the geometric viewpoint developed by Doyon *et al.* (2018).

Chapter 7

Linearized Hydrodynamics and GGE Spacetime Correlations

In equilibrium statistical mechanics, a central theme is to understand the structure of static correlations, with particular focus on critical points in thermodynamic parameter space in the neighborhood of which the correlation length is large in microscopic units. Natural extensions are time-dependent correlations, which can be viewed as the propagation of initially small perturbations in the equilibrium state. Now, conservation laws and broken symmetries will play a crucial role. The most elementary approach is the Landau–Lifshitz theory which uses the link to macroscopic equations linearized at thermal equilibrium, as is discussed in the following. Near critical points the more sophisticated techniques of critical dynamics would come into play.

7.1 Equilibrium spacetime correlations for nonintegrable chains

The general structure behind the Landau–Lifshitz theory can be explained already in the context of nonintegrable systems with a few conservation laws. For concreteness, we consider nonintegrable anharmonic chains. In essence, the extension to integrable chains consists in substituting the respective 3×3 matrices by operators on Hilbert spaces with infinite dimension. To distinguish from the infinite dimensional case, we use $\vec{u}$ for 3-vectors and A for 3×3 matrices.

The infinitely extended chain is governed by the Hamiltonian

$$H_{\mathrm{ch}} = \sum_{j \in \mathbb{Z}} \left(\tfrac{1}{2} p_j^2 + V_{\mathrm{ch}}(r_j) \right) \tag{7.1}$$

with equations of motion

$$\frac{d}{dt} r_j = p_{j+1} - p_j\,, \qquad \frac{d}{dt} p_j = V'_{\text{ch}}(r_j) - V'_{\text{ch}}(r_{j-1}). \tag{7.2}$$

The chain potential, V_{ch}, is bounded from below and increases at least one-sided linearly at infinity so as to have a finite partition function. Stretch, momentum, and energy are defined by the local fields

$$\vec{Q}_j = (r_j, p_j, 2e_j)\,, \quad e_j = \tfrac{1}{2}\big(p_j^2 + V_{\text{ch}}(r_{j-1}) + V_{\text{ch}}(r_j)\big), \tag{7.3}$$

and the respective current densities are

$$\vec{J}_j = \big(-p_j, -V'_{\text{ch}}(r_{j-1}), \quad -(p_{j-1} + p_j)V'_{\text{ch}}(r_{j-1})\big). \tag{7.4}$$

It is assumed that there are no further local conservation laws, which is the essence of nonintegrability.

The canonical equilibrium state factorizes with one factor given by

$$\frac{1}{Z_0(P, \bar{u}, \beta)} \exp\!\big(- \beta\big(\tfrac{1}{2}(p_0 - \bar{u})^2 + V_{\text{ch}}(r_0)\big) - Pr_0\big), \tag{7.5}$$

where we introduced the intensive dual parameters $P, \bar{u}, \beta$. $\mathfrak{p} = P/\beta$ is the physical pressure, $\bar{u}$ is the mean velocity, and β is the inverse temperature. The chain free energy per site equals

$$F_{\text{ch}} = -\log Z_0(P, \bar{u}, \beta). \tag{7.6}$$

Of course, more explicit expressions could be provided, but this is not needed at the moment. In accordance with conventions for the Toda chain, the indices run over $m, n = 0, 1, 2$, throughout. The static field–field correlator in infinite volume is defined by

$$C_{m,n}(j) = \langle Q_j^{[m]} Q_0^{[n]} \rangle^{\mathrm{c}}_{P,\bar{u},\beta} = \delta_{0j} \langle Q_0^{[m]} Q_0^{[n]} \rangle^{\mathrm{c}}_{P,\bar{u},\beta} \tag{7.7}$$

and the static field–field susceptibility matrix by

$$C_{m,n} = \sum_{j \in \mathbb{Z}} C_{m,n}(j) = \langle Q_0^{[m]} Q_0^{[n]} \rangle^{\mathrm{c}}_{P,\bar{u},\beta}. \tag{7.8}$$

C is the matrix of second derivatives of the chain free energy F_{ch}. In the same fashion, the field–current correlator is given by

$$B_{m,n}(j) = \langle J_j^{[m]} Q_0^{[n]} \rangle^{\mathrm{c}}_{P,\bar{u},\beta} \tag{7.9}$$

and the static field–current susceptibility matrix by

$$B_{m,n} = \sum_{j \in \mathbb{Z}} B_{m,n}(j). \tag{7.10}$$

The field–field spacetime correlator is the matrix

$$S_{m,n}(j,t) = \langle Q_j^{[m]}(t) Q_0^{[n]}(0) \rangle_{P,\bar{u},\beta}^{c}, \tag{7.11}$$

which for $t \neq 0$ is no longer δ-correlated in j. It is convenient to also introduce its Fourier transform

$$\hat{S}_{m,n}(k,t) = \sum_{j \in \mathbb{Z}} e^{ikj} S_{m,n}(j,t) \tag{7.12}$$

with $k \in [-\pi, \pi]$.

Switching to the macroscopic continuum scale, the average fields and currents are denoted by $\vec{u} = \langle \vec{Q}_0 \rangle_{P,\bar{u},\beta}$, $\vec{j} = \langle \vec{J}_0 \rangle_{P,\bar{u},\beta}$. According to the rules of thermodynamics, there is a one-to-one map between $\vec{u}$ and $(P,\bar{u},\beta)$. Therefore, the hydrodynamic equations for the chain can be written as

$$\partial_t \vec{u}(x,t) + \partial_x \vec{j}(\vec{u}(x,t)) = 0 \tag{7.13}$$

with quasilinear form

$$\partial_t \vec{u}(x,t) + A(\vec{u}(x,t)) \partial_x \vec{u}(x,t) = 0. \tag{7.14}$$

The matrix A is known as *flux Jacobian* and given through

$$A_{m,n}(\vec{u}) = \partial_{u_n} \langle J_0^{[m]} \rangle_{\vec{u}}, \tag{7.15}$$

which by the chain rule yields

$$A = BC^{-1}, \tag{7.16}$$

viewed as an identity in $(P,\bar{u},\beta)$. The flux Jacobian A is not symmetric in general. As established in (6.8), essentially using only spacetime stationarity, the matrix B turns out to be symmetric. By construction, C is symmetric and also $C > 0$. Thus,

$$A = C^{\frac{1}{2}} C^{-\frac{1}{2}} B^{-1} C^{-\frac{1}{2}} C^{-\frac{1}{2}}, \tag{7.17}$$

which implies that A has real eigenvalues, c_α, $\alpha = 0, 1, 2$, and a complete system of right, $|\psi_\alpha\rangle$, and left eigenvectors, $\langle \tilde{\psi}_\alpha |$.

Following Landau and Lifshitz, the average (7.11) is viewed as a small initial perturbation of equilibrium. Then to leading order, the correlator may be approximated by Eq. (7.14) linearized around a constant background with parameters $P, \bar{u}, \beta$. Denoting this perturbation by $\vec{u}$, the nonlinear hydrodynamic equation turns to its linearized version

$$\partial_t \vec{u}(x,t) + A \partial_x \vec{u}(x,t) = 0, \tag{7.18}$$

where the flux Jacobian A is now evaluated at the background, thus constant in spacetime. However, the initial conditions, $\vec{u}(x)$, are random and their average will be denoted by $\mathbb{E}$. According to thermal equilibrium, because of independence, on the macroscopic scale, the random field $\vec{u}(x)$ is Gaussian white noise with mean zero and covariance

$$\mathbb{E}\big(u_m(x)u_n(0)\big) = C_{m,n}\delta(x) = S_{m,n}^\sharp(x,0), \tag{7.19}$$

which defines the continuum static correlator $S^\sharp(x,0)$. To obtain its spacetime version, one has to solve (7.18) with initial conditions (7.19), which yields

$$\mathbb{E}\big(u_m(x,t)u_n(0,0)\big) = S_{m,n}^\sharp(x,t), \tag{7.20}$$

thereby defining the right-hand side. In Fourier space,

$$\hat{S}_{m,n}^\sharp(k,t) = (e^{-ikAt}C)_{m,n}. \tag{7.21}$$

Using the eigenvectors of A, in position space, the correlator (7.20) has the explicit form

$$S_{m,n}^\sharp(x,t) = \sum_{\alpha=0}^{2} \delta(x - c_\alpha t)(|\psi_\alpha\rangle\langle\tilde{\psi}_\alpha|C)_{m,n}. \tag{7.22}$$

To establish the relation to the microscopic definition (7.11), one introduces the dimensionless scale parameter ϵ. Then, under ballistic scaling, it is expected that for small ϵ,

$$\epsilon^{-1}S_{m,n}(\lfloor\epsilon^{-1}x\rfloor, \epsilon^{-1}t) \simeq S_{m,n}^\sharp(x,t). \tag{7.23}$$

Here, $\lfloor\cdot\rfloor$ denotes the integer part and the prefactor is chosen such that the sum rule

$$\sum_{j\in\mathbb{Z}} S(j,t) = \sum_{j\in\mathbb{Z}} S(j,0) \tag{7.24}$$

holds. In Fourier space, more compactly,

$$\lim_{\epsilon\to 0} \hat{S}_{m,n}(\epsilon k, \epsilon^{-1}t) = \hat{S}_{m,n}^\sharp(k,t). \tag{7.25}$$

The eigenvectors of A are called normal modes. The αth mode travels with velocity c_α and the initial condition determines the particular linear combination of normal modes as encoded by the susceptibility matrix C. The spacetime correlator has three delta peaks moving ballistically. If the background has zero average momentum, then the eigenvalues are $\vec{c} = (-c, 0, c)$ with c the isentropic speed of sound. There are two sound peaks with equal speed moving in opposite directions and a heat peak standing still. For an integrable system, there are so to speak infinitely many peaks, each moving with its own velocity. We thus expect the corresponding time correlator to have a broad spectrum which expands ballistically.

Actually, the peak structure on the Euler scale is only the starting step of the Landau–Lifshitz theory. Its main focus is the broadening of the peaks due to dissipation and molecular noise. For a simple fluid in three dimensions, the broadening is of order $\sqrt{t}$ with a Gaussian shape function. In one dimension, the standard Landau–Lifshitz theory would make the same prediction. But the linear fluctuation theory has to be corrected by expanding the nonlinear Euler term to second order. In Figure 7.1 shown is molecular dynamics of a fluid with shoulder interaction potential. Clearly

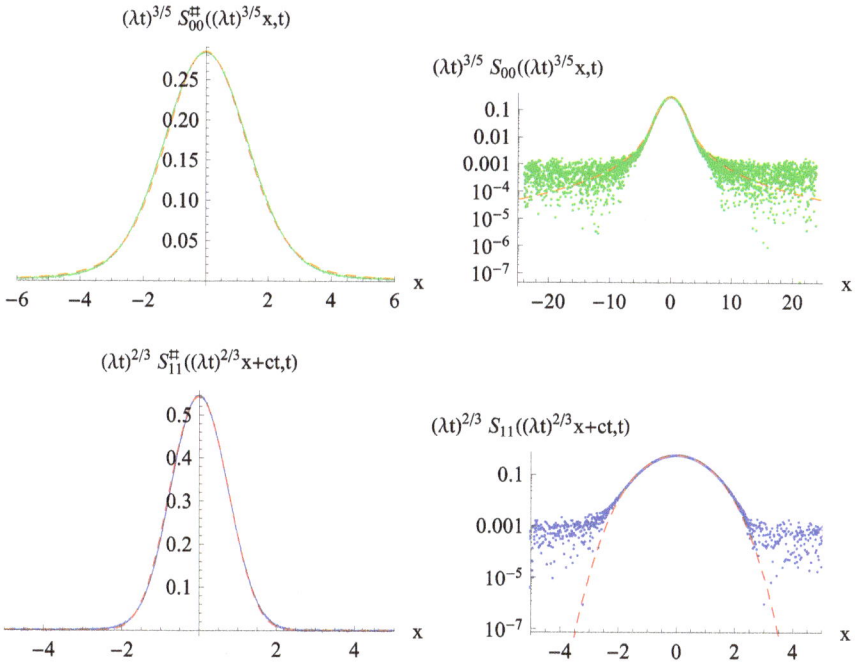

Fig. 7.1. Heat and sound peak of the equilibrium spacetime correlations for a fluid interacting through a shoulder potential. Shoulder means $V_{\mathrm{sh}}(x) = \infty$ for $|x| < \frac{1}{2}$, $V_{\mathrm{sh}}(x) = 1$ for $\frac{1}{2} < |x| < 1$, and $V_{\mathrm{sh}}(x) = 0$ for $1 < |x|$. Heat peak is on top and sound peak, displaced by ct, below. System size is $N = 4096$, time $t = 1024$, $\beta = 2$, and $P = 1.2$ for heat respectively $P = 1$ for sound. The sound speed is $c = 1.74$. There is a single scale-parameter in comparison with theoretical predictions. The right panels display the corresponding logarithmic plot. Numerically, the shoulder potential has the simplification that particle trajectories are piecewise linear. From Mendl and Spohn (2014).

visible is the anomalous broadening of the delta peaks with power laws $t^{3/5}$ and $t^{2/3}$. The shape functions are computed from the nonlinear version of the Landau–Lifshitz equations.

7.2 GGE spacetime correlations for the Toda lattice

For the Toda chain, we follow step by step the road map provided by the finite mode case, where we recall that $\varsigma_n(w) = w^n$, including $n = 0$. The field–field and field–current correlators have been introduced before in (6.5), resp. (6.6). Since we have computed already the GGE average of fields in (3.62) and of currents in (6.16), the matrix elements of C, B can be determined by differentiating these averages with respect to P and μ_n. The computation is somewhat lengthy and stated is merely the final result. For the field–field correlator, one obtains

$$C_{0,0} = \nu^3 \langle \rho_{\mathsf{p}} \varsigma_0^{\mathrm{dr}} \varsigma_0^{\mathrm{dr}} \rangle,$$

$$C_{0,n} = C_{n,0} = -\nu^2 \langle \rho_{\mathsf{p}} \varsigma_0^{\mathrm{dr}} (\varsigma_n - q_n \varsigma_0)^{\mathrm{dr}} \rangle,$$

$$C_{m,n} = \nu \langle \rho_{\mathsf{p}} (\varsigma_m - q_m \varsigma_0)^{\mathrm{dr}} (\varsigma_n - q_n \varsigma_0)^{\mathrm{dr}} \rangle, \qquad (7.26)$$

$m, n \geqslant 1$, and for the field–current correlator,

$$B_{0,0} = \nu^2 \langle \rho_{\mathsf{p}} (v^{\mathrm{eff}} - q_1) \varsigma_0^{\mathrm{dr}} \varsigma_0^{\mathrm{dr}} \rangle,$$

$$B_{0,n} = B_{n,0} = -\nu \langle \rho_{\mathsf{p}} (v^{\mathrm{eff}} - q_1) \varsigma_0^{\mathrm{dr}} (\varsigma_n - q_n \varsigma_0)^{\mathrm{dr}} \rangle,$$

$$B_{m,n} = \langle \rho_{\mathsf{p}} (v^{\mathrm{eff}} - q_1)(\varsigma_m - q_m \varsigma_0)^{\mathrm{dr}} (\varsigma_n - q_n \varsigma_0)^{\mathrm{dr}} \rangle. \qquad (7.27)$$

Both matrices have a two-block structure. Rather than using the discrete index, $n = 1, 2, \ldots$, we introduce some real-valued function, ϕ. Our linear space now consists of two-vectors (r, ϕ) with r the coefficient for the index 0 and ϕ for all other modes. More formally, the appropriate linear space is $\mathbb{C} \oplus \left(L^2(\mathbb{R}, \mathrm{d}w) \ominus \{\varsigma_0\} \right)$, the second summand consisting of all square-integrable functions orthogonal to ς_0. We also introduce the linear operator, F, in Dirac notation,

$$F = (1 - T\rho_{\mathsf{n}})^{-1} - \nu |\varsigma_0^{\mathrm{dr}}\rangle \langle \rho_{\mathsf{p}}|. \qquad (7.28)$$

Note that $F\varsigma_0 = 0$. Then,

$$C = \begin{pmatrix} \nu^3 \langle \rho_{\mathsf{p}} \varsigma_0^{\mathrm{dr}} \varsigma_0^{\mathrm{dr}} \rangle & -\nu^2 \langle \rho_{\mathsf{p}} \varsigma_0^{\mathrm{dr}} | F \\ -\nu^2 F^* | \rho_{\mathsf{p}} \varsigma_0^{\mathrm{dr}} \rangle & \nu F^* \rho_{\mathsf{p}} F \end{pmatrix}. \qquad (7.29)$$

As introduced before, for some general linear operator O, O^* stands for the adjoint operator of O with respect to the standard L^2 inner product $\langle \cdot, \cdot \rangle$.

In particular,

$$F^* = (1 - \rho_{\mathsf{n}}T)^{-1} - \nu|\rho_{\mathsf{p}}\rangle\langle\varsigma_0^{\mathrm{dr}}|. \tag{7.30}$$

With the same notation, the matrix B is given by

$$B = \begin{pmatrix} \nu^2\langle\rho_{\mathsf{p}}(v^{\mathrm{eff}} - q_1)\varsigma_0^{\mathrm{dr}}\varsigma_0^{\mathrm{dr}}\rangle & -\nu\langle\rho_{\mathsf{p}}(v^{\mathrm{eff}} - q_1)\varsigma_0^{\mathrm{dr}}|F \\ -\nu F^*|\rho_{\mathsf{p}}(v^{\mathrm{eff}} - q_1)\varsigma_0^{\mathrm{dr}}\rangle & F^*\rho_{\mathsf{p}}(v^{\mathrm{eff}} - q_1)F \end{pmatrix}. \tag{7.31}$$

Following the road map, we are supposed to compute $e^{At}C$ with $A = BC^{-1}$, which does not seem to be completely straightforward. But instead one might guess the entire solution $S(x,t)$ by noting that $S(x,0) = \delta(x)C$ and $\partial_t S(x,t)\big|_{t=0} = -\delta'(x)B$. Indeed, the two conditions can be satisfied by setting

$$S(x,t) = \begin{pmatrix} \nu^3\langle\rho_{\mathsf{p}}\,\delta(\Sigma(x,t))\varsigma_0^{\mathrm{dr}}\varsigma_0^{\mathrm{dr}}\rangle & -\nu^2\langle\rho_{\mathsf{p}}\,\delta(\Sigma(x,t))\varsigma_0^{\mathrm{dr}}|F \\ -\nu^2 F^*|\rho_{\mathsf{p}}\,\delta(\Sigma(x,t))\varsigma_0^{\mathrm{dr}}\rangle & \nu F^*\rho_{\mathsf{p}}\delta(\Sigma(x,t))F \end{pmatrix}, \tag{7.32}$$

with the shorthand

$$\Sigma(x,t) = x - t\nu^{-1}(v^{\mathrm{eff}} - q_1). \tag{7.33}$$

As a control check, the correlator scales indeed ballistically,

$$S(x,t) = t^{-1}S(x/t,1). \tag{7.34}$$

Our computation misses that matching simply the value and first derivative at $t = 0$ does not determine the full solution. Leaving for the moment a more convincing argument aside, let us reflect on the resulting predictions. For example, within the stated approximations, the spacetime stretch–stretch correlation function is predicted as

$$S_{0,0}(x,t) = \nu^2 \int_{\mathbb{R}} \mathrm{d}w\delta(x - t\nu^{-1}(v^{\mathrm{eff}}(w) - q_1))\nu\rho_{\mathsf{p}}(w)\varsigma_0^{\mathrm{dr}}(w)^2, \tag{7.35}$$

and similarly for the momentum–momentum correlation,

$$S_{1,1}(x,t) = \int_{\mathbb{R}} \mathrm{d}w\delta(x - t\nu^{-1}(v^{\mathrm{eff}}(w) - q_1))\nu\rho_{\mathsf{p}}(w)(\varsigma_1 - q_1\varsigma_0)^{\mathrm{dr}}(w)^2. \tag{7.36}$$

Both identities can be deduced most easily by going back to (7.26) upon inserting the δ-function. Our result is in perfect analogy to the finite mode case. The modes are labeled by the spectral parameter w and propagate with velocity $\nu^{-1}(v^{\mathrm{eff}}(w) - q_1)$. The last factor provides the weights and depends on the particular choice for the matrix elements of $S(x,t)$.

To confirm the guess, one has to figure out the spectral representation of the operator $A = BC^{-1}$. As key observation, in (6.24), we already introduced a nonlinear map which transforms the system of conservation laws into its quasilinear version in such a way that the operator corresponding to A is manifestly diagonal. This property is expected to persist when linearizing the map (6.24). In the ν, ρ_p variables, the map is given by

$$\rho_\mathsf{n} = \frac{\rho_\mathsf{p}}{1 + T\rho_\mathsf{p}}. \tag{7.37}$$

Both sides are linearized as $\rho_\mathsf{n} + \epsilon g$, $\nu + \epsilon r$, $\nu\rho_\mathsf{p} + \epsilon\phi$, $\langle\phi\rangle = 0$. To first order in ϵ, this yields the linear map $R : g \mapsto (r, \phi)$ given by

$$Rg = \nu \begin{pmatrix} -\nu\langle g(\varsigma_0^{\mathrm{dr}})^2\rangle \\ F^*(g\varsigma_0^{\mathrm{dr}}) \end{pmatrix} \tag{7.38}$$

with F^* as in (7.30). Note that indeed $\langle F^*\phi\rangle = 0$. Equation (7.37) can be inverted as

$$\rho_\mathsf{p} = (1 - \rho_\mathsf{n}T)^{-1}\rho_\mathsf{n}, \tag{7.39}$$

thereby deriving, by a similar argument as before,

$$R^{-1}\begin{pmatrix} r \\ \phi \end{pmatrix} = (\nu\varsigma_0^{\mathrm{dr}})^{-1}(-\rho_\mathsf{n}r + (1 - \rho_\mathsf{n}T)\phi). \tag{7.40}$$

Indeed, one checks that

$$RR^{-1} = 1, \quad R^{-1}R = 1, \tag{7.41}$$

the first "1" standing for the identity operator as a 2×2 block matrix and the second "1" for the identity operator in the space of scalar functions.

The operators C, B can be written in the new basis with the result

$$R^{-1}CR = \nu|\rho_\mathsf{p}\rangle\langle(\varsigma_0^{\mathrm{dr}})^2| + \nu(\varsigma_0^{\mathrm{dr}})^{-1}\rho_\mathsf{p}FF^*\varsigma_0^{\mathrm{dr}} \tag{7.42}$$

and

$$R^{-1}BR = |(v^{\mathrm{eff}} - q_1)\rho_\mathsf{p}\rangle\langle(\varsigma_0^{\mathrm{dr}})^2| + (v^{\mathrm{eff}} - q_1)(\varsigma_0^{\mathrm{dr}})^{-1}\rho_\mathsf{p}FF^*\varsigma_0^{\mathrm{dr}})$$
$$= \nu^{-1}(v^{\mathrm{eff}} - q_1)R^{-1}CR. \tag{7.43}$$

As anticipated, in the new basis $R^{-1}AR$ is simply multiplication by $\nu^{-1}(v^{\mathrm{eff}} - q_1)$. We conclude that

$$e^{At}C = R\exp(\nu^{-1}(v^{\mathrm{eff}} - q_1)t)R^{-1}C. \tag{7.44}$$

Working out the algebra, one arrives at

$$
\mathrm{e}^{At}C
$$
$$
= \begin{pmatrix} \nu^3\langle\rho_{\mathrm{p}}\exp\!\big(\nu^{-1}(v^{\mathrm{eff}}-q_1)t\big)(\varsigma_0^{\mathrm{dr}})^2\rangle & -\langle\rho_{\mathrm{p}}\exp\!\big(\nu^{-1}(v^{\mathrm{eff}}-q_1)t\big)\varsigma_0^{\mathrm{dr}}|F\rangle \\ -\nu^2 F^*|\rho_{\mathrm{p}}\exp\!\big(\nu^{-1}(v^{\mathrm{eff}}-q_1)t\big)\varsigma_0^{\mathrm{dr}}\rangle & \nu F^*\rho_{\mathrm{p}}\exp\!\big(\nu^{-1}(v^{\mathrm{eff}}-q_1)t\big)F \end{pmatrix}.
$$
$$(7.45)$$

Adding the spatial dependence, first in Fourier space, the propagator $\mathrm{e}^{At}C$ is modified to $\mathrm{e}^{\mathrm{i}kAt}C$, which in position space yields (7.32). For completeness, the matrix A is recorded as

$$
A = \nu^{-1}R(v^{\mathrm{eff}}-q_1)R^{-1}
$$
$$
= \begin{pmatrix} \langle\rho_{\mathrm{n}}(v^{\mathrm{eff}}-q_1)\varsigma_0^{\mathrm{dr}}\rangle & -\langle(v^{\mathrm{eff}}-q_1)\varsigma_0^{\mathrm{dr}}(1-\rho_{\mathrm{n}}T)| \\ -\nu^{-1}F^*|\rho_{\mathrm{n}}(v^{\mathrm{eff}}-q_1)\rangle & \nu^{-1}F^*(v^{\mathrm{eff}}-q_1)(1-\rho_{\mathrm{n}}T) \end{pmatrix}.
$$
$$(7.46)$$

In Figure 7.2, we report on molecular dynamics simulations of the Toda lattice in thermal equilibrium for three distinct parameter values. While there is a large body of work in providing microscopic confirmation of GHD in the context of a variety of models, presumably the data in Figure 7.2 constitute the most accurate check, so far. There are several points in favor. The initial statistical state is sharply defined. Modulo numerically computing the quantities appearing in the TBA formalism, there is the concrete prediction (7.32) based on linearized GHD. The comparison uses *no* adjustable parameter. Errors are smaller than 3.5%, which is of the order of statistical noise. For the shorter time $t = 150$ (data not displayed), one still observes ballistic scaling for our range of parameters. However, as systematic deviation, the peaks are slightly lower than predicted by the theory.

At the highest available pressure, the peaks are shifted toward the boundaries, which is even more pronounced for $\beta = 2.0, P = 3.53$ (data not displayed). This behavior reflects that for $\beta, P \to \infty$, at fixed ratio, the Toda DOS converges to the arcsine law which exhibits an inverse square root singularity at either boundary, compare with (3.75). To be emphasized is the middle frame at which $\nu = -0.03$, hence $|\rho_{\mathrm{f}}| = 33$. Closer to $\nu = 0$ the TBA simulation becomes difficult. Of course for molecular dynamics one would merely have to set $q_1 = q_N$ at which point the typical distance of particles is order $1/\sqrt{N}$, which equals 0.02 in our case. At such high densities, the collision rate ansatz becomes dubious. But the simulation confirms that GHD remains valid even under extreme conditions.

In principle, as for nonintegrable systems, there should be also dissipative broadening. As explained in Section 15.2, there are expected to be

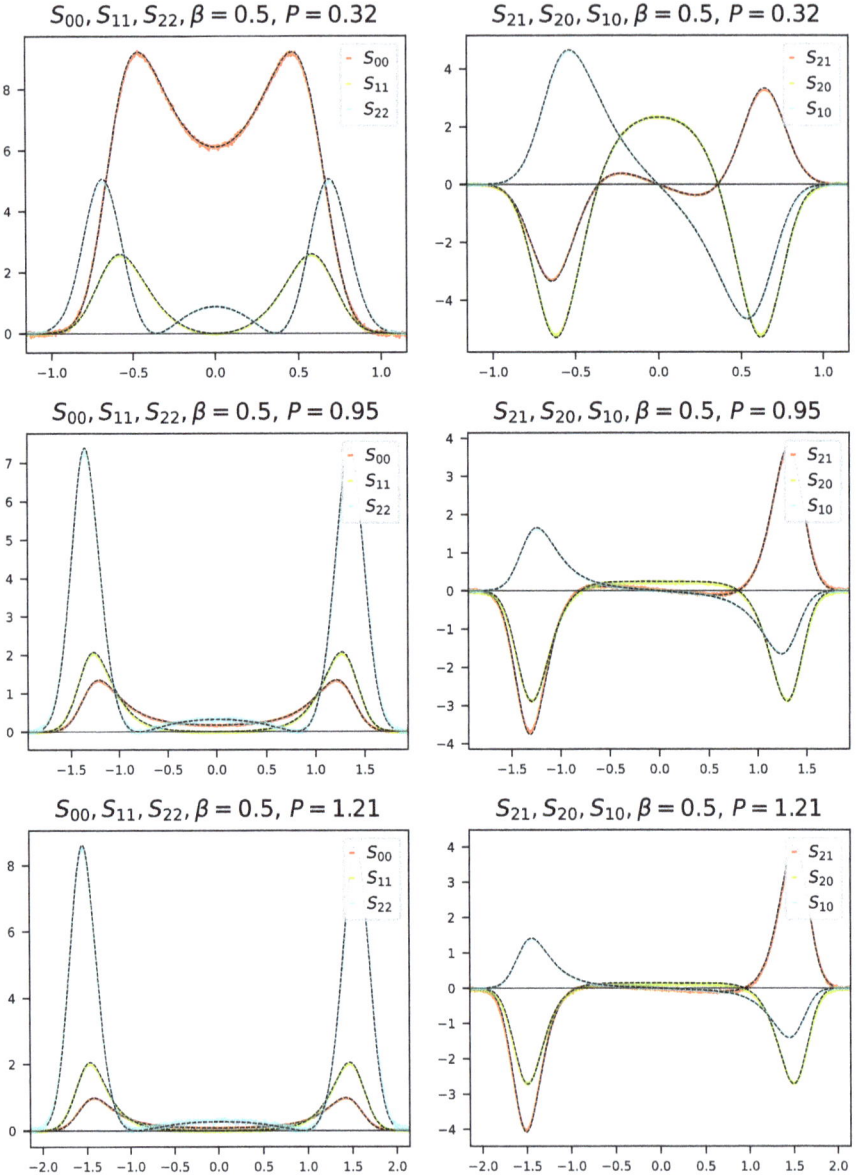

Fig. 7.2. Toda spacetime correlation functions in thermal equilibrium. The dashed line is the GHD prediction (7.32) for $m, n = 0, 1, 2$ and the colored lines are molecular dynamics simulations. The system size is $N = 3,000$ and time $t = 600$. The parameters β, P are indicated and correspond to $\nu = 2.58, -0.03, -0.42$. The thermal average is over samples is 3×10^6. From Mazzuca *et al.* (2023).

diffusive. To observe them quantitatively is blocked a generic obstacle. For a well-isolated peak, the sub-ballistic broadening can be easily observed. But for the Toda lattice, the Euler term yields already a smeared out background. So, one has to specifically tune to physical situations for which the anticipated dissipative corrections can be detected.

⚬⚬ *Equilibrium spacetime correlations for hard rods*: For hard rods, the T operator can be inverted and identities as (7.32) become more explicit. We consider thermal equilibrium, hence $h(w) = h_\beta(w) = \sqrt{\beta/2\pi}$ $\exp(-\beta w^2/2)$, $q_1 = 0$, and

$$v^{\mathrm{eff}}(v) = \frac{\nu}{\nu - a} v. \tag{7.47}$$

The dressing operator reads

$$(1 - T\rho_{\mathrm{n}})^{-1} = 1 + \frac{a}{\nu}|\varsigma_0\rangle\langle h_\beta| \tag{7.48}$$

and hence

$$\varsigma_0^{\mathrm{dr}} = 1 + \frac{a}{\nu}, \quad \varsigma_1^{\mathrm{dr}} = v, \quad \varsigma_2^{\mathrm{dr}} = v^2 + \frac{a}{\beta\nu}. \tag{7.49}$$

Rescaling the spatial coordinate to $x = (\nu - a)x/t$, one concludes

$$\{S_{00}^{\mathrm{hr}}(x,t), S_{11}^{\mathrm{hr}}(x,t), S_{22}^{\mathrm{hr}}(x,t)\}$$
$$= (\nu - a)t^{-1}h_\beta(x)\{1 + (a/\nu))^2, x^2, (x^2 + (a/\beta\nu))^2\}. \tag{7.50}$$

The interesting term is the energy–energy spacetime correlation $S_{22}^{\mathrm{hr}}(x,t)$. For the ideal gas, $a = 0$, the scaling function $x^4 h_\beta(x)$ vanishes at $x = 0$ and has a double peak. Turning on the interaction, the minimum at $x = 0$ is shifted to $h_\beta(0)(a/\beta\nu)^4$ as local maximum, resulting in three peaks.   ⚬⚬

Notes and references

Section 7.0

The book by Forster (1975) is still a very readable exposition of the fluctuation theory developed by Landau and Lifshitz (1975). Critical dynamics is covered in the classic review of Hohenberg and Halperin (1977).

Section 7.1

The existence of the infinite volume dynamics is proved by Lanford *et al.* (1977). Fluctuation theory for simple fluids is discussed in Spohn (1991). As a crucial distinction, the hydrodynamic equations for a few conservation laws may develop shocks. An instructive account is the most readable lecture course by Bressan (2013). In Das *et al.* (2020), a spin chain is considered, for which at low temperatures phase differences emerge as an almost conserved field, on top of the two strictly conserved fields.

Section 7.2

A more detailed account of GGE spacetime correlations can be found in Spohn (2020). Early molecular dynamics simulations of the Toda lattice and related theoretical investigations have been carried out by Schneider and Stoll (1980), Schneider (1983), Diederich (1981), and Cuccoli *et al.* (1993). Molecular dynamics simulations of thermal spacetime correlations are reported by Kundu and Dhar (2016). Among other features, investigated are the low-density regime and the low-temperature harmonic approximation. A wider range of parameters is covered by Mazzuca *et al.* (2023) including self- and cross-correlations of stretch, momentum, and energy. The numerical results are compared with the GHD prediction in Eq. (7.32). A more general perspective on fluctuations at the Euler scale and beyond is developed by Doyon and Myers (2020).

Chapter 8

Domain Wall Initial States

Spacetime correlations encode small deviations from a GGE. But the hydrodynamic scale covers also initial states very far from stationarity, thereby exhibiting a rich portfolio of time-dependent behavior. From a theoretical perspective, a very natural and much studied nonequilibrium initial state is enforced by joining two semi-infinite systems: the left half line in one GGE and the right half line in another GGE. The fields are spatially constant except for a single jump at the origin. On the level of GHD, the solution scales exactly as x/t, a simplification which promises to yield exact solutions. In the mathematical theory of finitely many hyperbolic conservation laws, such a set-up is known as Riemann problem. Its solution consists of flat pieces, smooth rarefaction waves, and jump discontinuities, called shocks. Their precise spatial sequence can be complicated and is, in principle, encoded by the eigenvectors and eigenvalues of the matrix A in their dependence on the thermodynamic parameters, compare with (7.15). As discussed in Section 6.2, in normal form, the flux Jacobian A is multiplication by $\nu^{-1}(v^{\text{eff}} - q_1)$. This form admits only rarefaction waves. No spontaneous generation of jump discontinuities is expected.

Besides its intrinsic interest, we pick our example to also illustrate the predictive power of GHD. For the domain wall problem, the primary goal is to figure out macroscopic behavior, i.e., how stretch, momentum, and energy vary spatially given their particular initial jump discontinuity. Apparently, GHD is the only theoretical scheme through which such a macroscopic behavior can be computed.

We slightly rewrite Eq. (6.25) as

$$\partial_t \rho_{\mathsf{n}}(x, t; v) + \tilde{v}^{\text{eff}}(x, t; v)\partial_x \rho_{\mathsf{n}}(x, t; v) = 0 \qquad (8.1)$$

with

$$\tilde{v}^{\text{eff}}(v) = \nu^{-1}(v^{\text{eff}}(v) - q_1). \tag{8.2}$$

Domain wall initial conditions read

$$\rho_{\text{dw}}(x, 0; v) = \chi(\{x < 0\})\rho_{\text{n}-}(v) + \chi(\{x \geqslant 0\})\rho_{\text{n}+}(v). \tag{8.3}$$

Instead of $\rho_{\text{n}\pm}$, physically, it might be more natural to prescribe the DOS of the Lax matrix and the average stretch. But mathematically, the normal form (8.1) together with (8.3) is more accessible.

Since the solution to (8.1), (8.3) scales ballistically, we set $\rho_{\text{dw}}(x, t; v) = \mathbf{g}(t^{-1}x; v)$ and $\tilde{v}^{\text{eff}}(x, t; v) = \tilde{v}^{\text{eff}}(t^{-1}x; v)$. Without loss of generality, one adopts $t = 1$ and arrives at

$$(x - \tilde{v}^{\text{eff}}(x; v))\partial_x \mathbf{g}(x; v) = 0, \quad \lim_{x \to \pm\infty} \mathbf{g}(x; v) = \rho_{\text{n}\pm}(v). \tag{8.4}$$

Therefore, $x \mapsto \mathbf{g}(x; v)$ for fixed v has to be constant except for jumps at the zeros of $x - \tilde{v}^{\text{eff}}(x; v)$ as a function of x. At this stage, it is not clear which level of generality to adopt. Physically, one would expect to have a unique solution. Thus, in the (x, v)-plane, there should be a *contact line* which divides the plane into two domains: one characterized to contain the set $\{-\infty\} \times \mathbb{R}$ and the other the set $\{\infty\} \times \mathbb{R}$. The solution $\mathbf{g}(x; v)$ is constant in either domain and jumps from $\rho_{\text{n}-}(v)$ to $\rho_{\text{n}+}(v)$ across the contact line. Uniqueness means a unique contact point for every v. Then, the contact line is represented as the graph of a function defined by $v \mapsto \tilde{\phi}(v)$. As discussed in the following, for hard rods, $\tilde{\phi}$ is invertible with inverse denoted by ϕ. Hence, the contact line can also be written as the graph of $x \mapsto \phi(x)$. We present our argument for the latter case, since hard rods are reasonably close to the Toda lattice. More general contact lines could be handled in a similar fashion.

With our assumptions, for every x there is a unique contact point $v = \phi(x)$, which satisfies

$$x - \tilde{v}^{\text{eff}}(x; v), \tag{8.5}$$

and the solution ansatz reads

$$g^\phi(v) = \chi(\{v > \phi\})\rho_{\text{n}-}(v) + \chi(\{v \leqslant \phi\})\rho_{\text{n}+}(v). \tag{8.6}$$

The superscript ϕ will be used to generically indicate that in the TBA formalism ρ_{n} is replaced by g^ϕ. For example,

$$f^{\text{dr},\phi} = \left(1 - Tg^\phi\right)^{-1}f, \quad \nu^\phi \langle \rho_{\text{p}}^\phi \rangle = 1, \tag{8.7}$$

compare with (3.57) and (3.60). Note that $g^{\pm\infty}(v) = \rho_{n\pm}(v)$. In particularly, we define

$$\tilde{v}^{\text{eff},\phi}(v) = (\nu^\phi)^{-1}\left(\frac{\varsigma_1^\phi(v)}{\varsigma_0^\phi(v)} - q_1^\phi\right). \tag{8.8}$$

From the solution ansatz, it follows that

$$\mathbf{g}(x;v) = g^{\phi(x)}(v), \quad \tilde{v}^{\text{eff}}(x;v) = \tilde{v}^{\text{eff},\phi(x)}(v). \tag{8.9}$$

The condition (8.5) turns into

$$x = \tilde{v}^{\text{eff},\phi(x)}(\phi(x)). \tag{8.10}$$

Defining

$$G(\phi) = \tilde{v}^{\text{eff},\phi}(\phi), \tag{8.11}$$

we conclude that

$$x = G(\phi(x)), \quad \phi(x) = G^{-1}(x). \tag{8.12}$$

The contact line is the inverse of the function G in (8.11). To determine G numerically, one has to compute the dressing depending on ϕ. This then yields $\mathbf{g}(x;v)$ from which the observables of interest, as average stretch, momentum, and energy, at location x are deduced.

Qualitatively, G is linear for large ϕ. But this is a special feature of fluid-like models with a quadratic kinetic energy. For the discrete linear

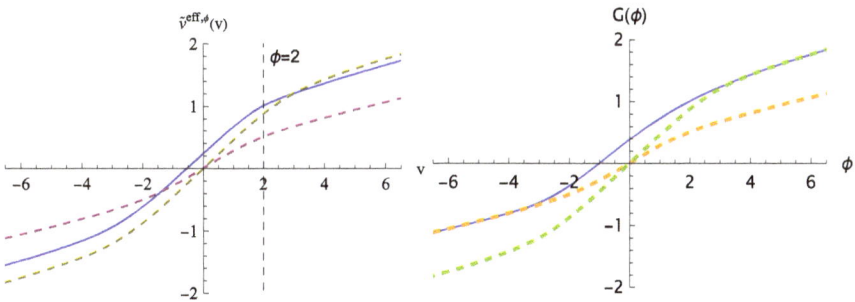

Fig. 8.1. Domain wall initial state for the Toda lattice at uniform inverse temperature $\beta = 1$, $P_- = \frac{1}{2}$, and $P_+ = 2$. Displayed in blue are $\tilde{v}^{\text{eff},\phi}(v)$ at $\phi = 2$ and the contact line $G(\phi)$. On the left, the dotted magenta line is $\tilde{v}^{\text{eff}}(v)$ computed by using ρ_{n+} and the olive line using ρ_{n-}. Correspondingly on the right displayed is $\tilde{v}^{\text{eff}}(\phi)$ using either ρ_{n+} or ρ_{n-}. From Mendl and Spohn (2022).

wave equation with a general dispersion relation, because of the bounded momentum space, the concept of contact line as such remains untouched, but the line is no longer equal to the graph of a function.

The predicted sharp step appears to be oversimplified. Indeed on longer time scales the step is expected to be smeared roughly as an error function, compare with Section 15.3.

◆◆ *Hard rod domain wall*: For the hard rod lattice, the function G can still be computed analytically. We recall that ν is the average stretch, $\nu > a$, and $h(v)$ denotes the normalized velocity distribution,

$$\rho_{\mathsf{p}}(v) = \nu^{-1}h(v), \quad \rho_{\mathsf{n}}(v) = (\nu - a)^{-1}h(v), \quad q_1 = \int_{\mathbb{R}} \mathrm{d}w\, h(w)w. \quad (8.13)$$

In normal form, the hard rod GHD reads

$$\partial_t \rho_{\mathsf{n}}(x,t;v) + (\rho(x,t) - a)^{-1}(v - q_1(x,t))\partial_x \rho_{\mathsf{n}}(x,t;v) = 0, \quad (8.14)$$

compare with the structurally identical Eq. (8.1). The initial condition is the same as in (8.3) and the G-function is computed by the method explained above with the result

$$G_{\mathrm{hr}}(\phi) = \left(\int_\phi^\infty \mathrm{d}v \rho_{\mathsf{n}-}(v) + \int_{-\infty}^\phi \mathrm{d}v \rho_{\mathsf{n}+}(v) \right)$$

$$\times \phi - \int_\phi^\infty \mathrm{d}v\, v\rho_{\mathsf{n}-}(v) - \int_{-\infty}^\phi \mathrm{d}v\, v\rho_{\mathsf{n}+}(v). \quad (8.15)$$

Clearly, $dG_{\mathrm{hr}}(\phi)/d\phi > 0$. Thus, G_{hr} is invertible. For $\phi \to \infty$, one obtains

$$G_{\mathrm{hr}}(\phi) \simeq \frac{1}{\nu_+ - a}(\phi - q_{1,+}) \quad (8.16)$$

and correspondingly for $\phi \to -\infty$ with ν_+, q_{1+} replaced by ν_-, q_{1-}. If the boundary velocity distributions have an exponential decay, then the error in (8.16) is also exponentially small. The contact line is asymptotically linear with a well-localized monotone interpolation.

If $a < 0$, one can tune the boundary values such that $\nu_- < 0 < \nu_+$. There is then some intermediate point, x_0 with $\nu(x_0) = 0$, a case studied in more detail at the end of Section 9.1.   ◆◆

Notes and references

Chapter 8

This chapter is based on the study of Mendl and Spohn (2022), where in particular numerical simulations of GHD with domain wall initial conditions are reported. Domain walls for hard rods are studied in Doyon and Spohn (2017). On the Euler scale, the step at the contact line is sharp. Earlier work of Bulchandani *et al.* (2019) considers the case of low pressure, $P_- < P_+ < P_c$. Molecular dynamics is compared with predictions of GHD. Also, quantum mechanical corrections are obtained to the lowest order. For the Toda lattice, diffusive corrections are discussed in Section 15.3. The broadening is convincingly observed in the XXZ model, see De Nardis *et al.* (2018).

Domain wall initial conditions have been studied numerically for the discrete sinh-Gordon model by Bastianello *et al.* (2018). The most detailed investigations are available for the XXZ model, see Piroli *et al.* (2017) and Misguich *et al.* (2017). In this case, the spectral parameter space contains in addition the type of string states and the contact line refers to a larger space. Gamayun *et al.* (2019) study domain wall initial conditions for a classical spin chain and its quantized version.

Due to momentum conservation, the contact line of the Toda lattice is one-to-one and covers the full real line. On the other hand, for XXZ, and other discrete models, the contact line takes values only in a bounded v-interval, which implies left and right edges in x-space up to which boundary values remain constant. The behavior near the edge often shows intricate oscillatory decay, which is beyond the hydrodynamic scale and has been elucidated in considerable detail for a variety of models, see Collura *et al.* (2018), Bulchandani and Karrasch (2019), and Grava *et al.* (2021).

Chapter 9

Toda Fluid

Toda particles might as well be viewed to move on the real line, which is the fluid picture, see Figure 2.1. The name "fluid" is somewhat misleading, since particles are distinguishable. Still, the dynamical behavior is fluid-like. In a way, we have to start from the beginning and redo the computation for average fields and their currents. Since we rely on chain results, the discussion can be more compressed. To avoid duplicating symbols, the fluid is distinguished from the chain through the index "$_\mathsf{f}$". $x \in \mathbb{R}$ stands for the coordinate of the one-dimensional physical space.

9.1 Euler equations

For the Toda fluid, conserved fields have a density given by

$$Q_\mathsf{f}^{[0]}(x) = \sum_{j \in \mathbb{Z}} \delta(q_j - x), \quad Q_\mathsf{f}^{[n]}(x) = \sum_{j \in \mathbb{Z}} \delta(q_j - x) Q_j^{[n]} \qquad (9.1)$$

for $n \geq 1$. Taking their time derivative,

$$\frac{\mathrm{d}}{\mathrm{d}t} Q_\mathsf{f}^{[0]}(x) = \sum_{j \in \mathbb{Z}} \delta'(q_j - x) p_j,$$

$$\frac{\mathrm{d}}{\mathrm{d}t} Q_\mathsf{f}^{[n]}(x) = \sum_{j \in \mathbb{Z}} \left(\delta'(q_j - x) p_j Q_j^{[n]} + \delta(q_j - x)(J_j^{[n]} - J_{j+1}^{[n]}) \right). \qquad (9.2)$$

Hence, fluid current densities read

$$J_\mathsf{f}^{[0]}(x) = \sum_{j \in \mathbb{Z}} \delta(q_j - x) p_j,$$

$$J_\mathsf{f}^{[n]}(x) = \sum_{j \in \mathbb{Z}} \left(\delta(q_j - x) p_j Q_j^{[n]} + \left(\theta(q_j - x) - \theta(q_{j-1} - x) \right) J_j^{[n]} \right) \qquad (9.3)$$

with θ the step function, $\theta(x) = 0$ for $x \leqslant 0$ and $\theta(x) = 1$ for $x > 0$. The index 0 has been treated separately. But, in fact, $Q_j^{[n]}$ naturally extends to $n = 0$ by setting $Q_j^{[0]} = 1$. This property foreshadows a single hydrodynamic equation.

The next item is GGE, where we start from the canonical ensemble with partition function

$$Z_{\text{can}}(N, \ell) = \int_{\Gamma_N} \prod_{j=1}^N dr_j dp_j \delta\big(Q^{[0],N} - \ell\big) \exp\big(- \text{tr}[V(L_N)]\big). \qquad (9.4)$$

The ensemble is canonical, since volume, $q_{N+1} - q_1$, and number of particles are fixed. In our discussion, the confining potential will play a passive role and is hence dropped from the notation, except when appearing as GGE index. For the chain, we considered $\ell = \nu N$ and, by the equivalence of ensembles, switched to $\exp\big(- PQ^{[0],N}\big)$ with the pressure P dual to the stretch ν. In the limit $N \to \infty$, the free energy per site, $F_{\text{to}}(P)$, has been obtained already, compare with (3.40). Turning to the fluid, we want to keep the volume fixed and allow for fluctuations in N. The corresponding chemical potential is denoted by μ. Then,

$$Z_{\text{f}}(\mu, \ell) = \sum_{N=1}^\infty e^{\mu N} \int_{\Gamma_N} \prod_{j=1}^N dr_j dp_j \delta\big(Q^{[0],N} - \ell\big) \exp\big(- \text{tr}[V(L_N)]\big), \quad (9.5)$$

compare with (3.1). One has to distinguish the cases $\ell > 0$ and $\ell < 0$, corresponding to either increasing or decreasing particle labeling, on average. Thus, the particle number $Q_{\text{f}}^{[0],N} > 0$, while the fluid density $\rho_{\text{f}} = N/\ell$ can take either sign. Since $\nu = \ell/N$, the relation $\nu\rho_{\text{f}} = 1$ holds always. Close to P_{c}, defined by $\nu(P_{\text{c}}) = 0$, the fluid density ρ_{f} jumps from ∞ to $-\infty$ when increasing P. This looks singular but merely reflects a particular choice of coordinates. $\nu(P)$ is a smooth function. Of interest is the limit

$$\lim_{\ell \to \pm\infty} -\frac{1}{\ell} \log Z_{\text{f}}(\mu, \ell) = F_{\text{f},\pm}(\mu), \qquad (9.6)$$

where $\pm$ refers to the sign of ℓ. The fluid free energy has two branches labeled by $\text{sgn}(\ell)$. Correspondingly, there are two distinct infinite volume GGE averages denoted by $\langle \cdot \rangle_{\mu,\pm,V}$. For a convergent partition function, one has to require $\mu < \mu_{\text{max}}$ for either branch with $\mu_{\text{max}} = F_{\text{to}}(P_{\text{c}})$, as displayed in Figure 9.1.

Thermodynamically, the dual to P is the stretch ν and to μ the fluid density ρ_{f}. Let us denote by $\mathcal{L}F$ the Legendre transform of some function F.

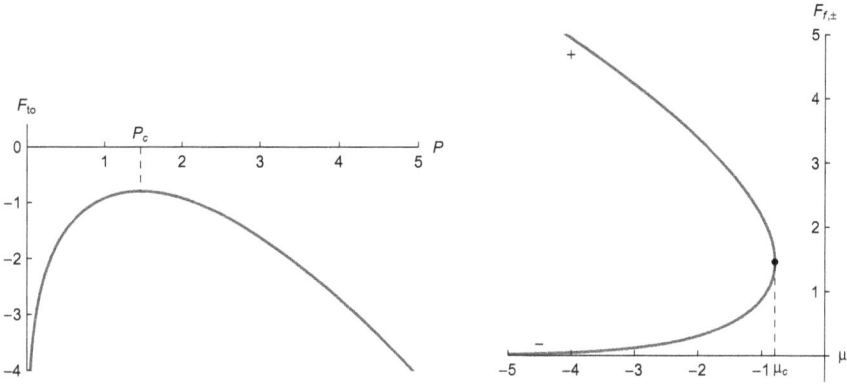

Fig. 9.1. For thermal equilibrium at $\beta = 1$, to the left is the Toda lattice free energy $F_{to} = \mu$ as a function of the pressure P. To the right are the two branches of the fluid free energy $F_{f,\pm} = -P$ as a function of μ, see (9.10). Courtesy of C. Mendl.

By the equivalence of ensembles on the level of free energies, one concludes that

$$\lim_{N \to \infty} -N^{-1} \log Z_{can}(N, \nu N) = \mathcal{L}F_{to}(\nu) \tag{9.7}$$

and

$$\lim_{\ell \to \pm\infty} -\ell^{-1} \log Z_{can}(\rho_f \ell, \ell) = \mathcal{L}F_{f,\pm}(\rho_f). \tag{9.8}$$

Thus, the two free energies are related by

$$\mathcal{L}F_{to}(\nu) = \rho_f^{-1} \mathcal{L}F_{f,\pm}(\rho_f), \quad \nu \rho_f = 1, \tag{9.9}$$

which implies

$$F_{to}(P) = \mu, \quad -F_{f,\pm}(\mu) = P. \tag{9.10}$$

Stated differently, F_{to} and $-F_{f,\pm}$ are inverse functions of each other. The first identity was already noted in (3.40) and the argument of $F_{f,\pm}$ is in fact the parameter appearing in the TBA equation. $F_{to}(P)$ is concave with maximum at P_c. On the other hand, $F_{f,+}$ is concave and $F_{f,-}$ convex with both graphs being smoothly joined at μ_{max}.

The free energy of the Toda fluid can be written in such a way as to suggest the extension to a larger class of integrable models. Starting from (9.10) and using (3.40), the free energy equals

$$F_f(\mu, V) = -\int_{\mathbb{R}} dw e^{-\varepsilon(w)}, \tag{9.11}$$

compare with (9.10) and (3.37). Differentiating the TBA equation with respect to μ, one arrives at

$$-\partial_\mu F_{\mathsf{f}}(\mu, V) = \int_{\mathbb{R}} dw \rho_{\mathsf{n}}(w)(1 - T\rho_{\mathsf{n}})^{-1} \varsigma_0(w) = \nu^{-1} = \rho_{\mathsf{f}}. \tag{9.12}$$

Similarly, by perturbing V as $V + \kappa \varsigma_n$, one concludes

$$\partial_\kappa F_{\mathsf{f}}(\mu, V + \kappa \varsigma_n)|_{\kappa=0} = \int_{\mathbb{R}} dw \rho_{\mathsf{n}}(w)(1 - T\rho_{\mathsf{n}})^{-1} \varsigma_n(w) = \langle \rho_{\mathsf{p}} \varsigma_n \rangle$$

$$= \langle Q_{\mathsf{f}}^{[n]}(0) \rangle_{\mu, V}, \tag{9.13}$$

which confirms the defining derivatives for the fluid free energy F_{f}.

For the current averages, we follow the strategy as developed for the lattice case by introducing the field–current susceptibility matrix through

$$B_{\mathsf{f}, m, n} = \int_{\mathbb{R}} dx \langle J_{\mathsf{f}}^{[m]}(x) Q_{\mathsf{f}}^{[n]}(0) \rangle_{\mu, \pm, V}^{\mathsf{c}}. \tag{9.14}$$

By the same argument as leading to (6.9), B_{f} is a symmetric matrix, $B_{\mathsf{f}, m, n} = B_{\mathsf{f}, n, m}$. Since $J_{\mathsf{f}}^{[0]} = Q_{\mathsf{f}}^{[1]}$, one starts from

$$\partial_\mu \langle J_{\mathsf{f}}^{[n]}(0) \rangle_{\mu, \pm, V} = B_{\mathsf{f}, n, 0} = B_{\mathsf{f}, 0, n} = \int_{\mathbb{R}} dx \langle Q_{\mathsf{f}}^{[1]}(x) Q_{\mathsf{f}}^{[n]}(0) \rangle_{\mu, \pm, V}^{\mathsf{c}}. \tag{9.15}$$

The last expression can be written as a derivative,

$$B_{\mathsf{f}, 0, n} = -\partial_\kappa \langle Q_{\mathsf{f}}^{[n]}(0) \rangle_{\mu, \pm, V + \kappa w}\big|_{\kappa=0} = -\partial_\kappa \langle \rho_{\mathsf{p}, \kappa} \varsigma_n \rangle\big|_{\kappa=0}, \tag{9.16}$$

where $\rho_{\mathsf{p}, \kappa}$ is determined by the solution of the TBA equation

$$V + \kappa \varsigma_1 - \mu - T\rho_{\mathsf{n}, \kappa} + \log \rho_{\mathsf{n}, \kappa} = 0. \tag{9.17}$$

In (3.60), we differentiated TBA with respect to ς_0. This time it is with respect to ς_1. Denoting by $'$ the derivative at $\kappa = 0$, one obtains

$$\rho_{\mathsf{n}}' = -\rho_{\mathsf{n}}(1 - T\rho_{\mathsf{n}})^{-1} \varsigma_1 \tag{9.18}$$

and

$$\rho_{\mathsf{p}}' = \partial_\mu \rho_{\mathsf{n}}' = -\partial_\mu (\rho_{\mathsf{n}} \varsigma_1^{\mathrm{dr}}) = -\partial_\mu (\rho_{\mathsf{p}} v^{\mathrm{eff}}), \tag{9.19}$$

where, as before, v^{eff} is defined in (6.20). Therefore,

$$\partial_\mu \left(\langle J_{\mathsf{f}}^{[n]}(0) \rangle_{\mu, \pm, V} - \langle \rho_{\mathsf{p}} v^{\mathrm{eff}} \varsigma_n \rangle \right) = 0, \tag{9.20}$$

implying that the difference does not depend on μ. To fix the constant, the current vanishes in the limit of vanishing density of particles, in other words, for $\mu \to -\infty$, and so does ρ_{p} since $\nu \langle \rho_{\mathsf{p}} \rangle = 1$. We conclude that

$$\langle J_{\mathsf{f}}^{[n]}(0) \rangle_{\mu, \pm, V} = \langle \rho_{\mathsf{p}} v^{\mathrm{eff}} \varsigma_n \rangle, \tag{9.21}$$

which holds for all n, including $n = 0$. Compared to the lattice currents (6.18), only the shift $\rho_{\mathsf{p}} q_1$ has been removed. This can be understood by starting from the fluid side. Then, switching to the lattice dynamical equations, the lateral motion is subtracted and such a term has to appear in the average lattice currents.

Since B_f is a symmetric matrix, there has to be a current potential, denoted by G_f such that

$$\partial_\kappa G_f(\mu, V + \kappa\varsigma_n)|_{\kappa=0} = \langle J_f^{[n]}(0)\rangle_{\mu,V}. \tag{9.22}$$

Indeed, one only has to slightly modify the free energy as

$$G_f(\mu, V) = -\int_{\mathbb{R}} dw w e^{-\varepsilon(w)}. \tag{9.23}$$

Then, as before by differentiating TBA,

$$\partial_\kappa G_f(\mu, V + \kappa\varsigma_n)|_{\kappa=0} = \int_{\mathbb{R}} dw w \rho_n(w)(1 - T\rho_n)^{-1}\varsigma_n(w) = \langle \varsigma_1^{dr}\rho_n\varsigma_n\rangle. \tag{9.24}$$

Since in (9.21) the term $n = 0$ no longer plays a special role, there is only a single hydrodynamic equation which reads

$$\partial_t \rho_p(x, t; v) + \partial_x\big(v^{eff}(x, t; v)\rho_p(x, t; v)\big) = 0. \tag{9.25}$$

Using (6.24) to transform to quasilinear form yields

$$\partial_t \rho_n(x, t; v) + v^{eff}(x, t; v)\partial_x \rho_n(x, t; v) = 0, \tag{9.26}$$

which is identical to the lattice version, except for the term accounting for the lateral motion. By construction, $\rho_n(x, t; v) \geqslant 0$, a property which is correctly propagated by (9.26). On the other hand, the overall sign of $\rho_p(x, t; v)$ is governed by $\nu(x, t)$, to say, if $\nu(x, t) > 0$, then $\rho_p(x, t; v) \geqslant 0$, and if $\nu(x, t) < 0$, then $\rho_p(x, t; v) \leqslant 0$. An example for $\nu(x, t)$ changing sign is discussed in the Insert just below.

◆◆ *High-low pressure domain wall*: As discussed in Chapter 8, a domain wall initial state consists of two half lines with homogeneous states ρ_{n-}, resp. ρ_{n+}. If $\nu_- < 0 < \nu_+$, then there is a sign change of the particle density ρ_p. While in Chapter 8 we used the Toda lattice, it is instructive to view the high-low pressure domain wall from the perspective of a fluid. To be concrete, in the initial state, the particle with label 0 is placed at the origin, $q_0 = 0$, the particles with negative indices correspond to a thermal state with parameters (ν_-, β) and those with positive indices to (ν_+, β). Thus, the momenta are i.i.d. Gaussian with mean zero and variance β^{-1}, while $\{q_{j+1} - q_j, j \leqslant -1\}$ are i.i.d. χ_{2P_-} distributed random variables, and $\{q_{j+1} - q_j, j \geqslant 0\}$ are i.i.d. χ_{2P_+}, compare with Chapter 8. The case of interest is $P_- > P_c > P_+$, thus $\nu_- < 0 < \nu_+$. Then, by assumption, $\langle q_{j+1} - q_j \rangle_{P_-} = \nu_- < 0$ for $j \leqslant -1$, while $\langle q_{j+1} - q_j \rangle_{P_+} = \nu_+ > 0$ for $j \geqslant 0$.

A typical ordering is of the form $\ldots > q_{-1} > q_0 = 0 < q_1 < \ldots$. Thus, initially, the average particle density equals $|\nu_-|^{-1} + \nu_+^{-1}$ on $[0, \infty)$ and decays exponentially on $(-\infty, 0]$. Under the dynamics, the point at which the ordered domain touches the anti-ordered domain is moving in time, its label being denoted by $\kappa(t)$. Close to $q_{\kappa(t)}$, particles pile up. The domain boundary acts as bottleneck for particles. The position of the bottleneck is $x_{\mathrm{bn}}(t) = q_{\kappa(t)}$. Numerical simulations suggest that $\kappa(t)$ and $x_{\mathrm{bn}}(t)$ change linearly in time, at least approximately. Also, to the left of the bottleneck, the particle density vanishes rapidly. As observed numerically, the scaling function G, see (8.12), has a non-zero slope at x_c. Thus, near the bottleneck, the particles should be distributed according to equilibrium with a linearly varying pressure. From this property, one infers that in physical space, the particle density at the bottleneck diverges as an inverse square root. It would be of interest to better understand the precise particle statistics close to the bottleneck. ◆◆

9.2 Generalized free energy — again

The fluid picture suggests an alternative approach to obtain the generalized free energy, thereby circumventing entirely the Dumitriu–Edelman change of volume elements. The same method will be applied to Calogero–Moser models which is discussed in Chapter 11. As further spin-off, a closer structural similarity to the Lieb–Liniger δ-Bose gas will be accomplished.

One starts from the Toda fluid, $(q_j, p_j) \in \mathbb{R}^2$, $j = 1, \ldots, N$. The open chain is governed by the Hamiltonian of Eq. (2.60) and has the Lax matrix

$$(L_N^\diamond)_{j,j} = p_j, \quad (L_N^\diamond)_{j,j+1} = (L_N^\diamond)_{j+1,j} = \mathrm{e}^{-((q_{j+1}-q_j)/2)} \tag{9.27}$$

with $j = 1, \ldots, N$ in the former case, $j = 1, \ldots, N - 1$ in the latter case, and $(L_N^\diamond)_{i,j} = 0$ otherwise. Some aspects of the scattering theory for the open Toda chain have been discussed already in Section 2.3. The explicit canonical transformation to scattering coordinates, alias action–angle variables, is stated in Eqs. (2.68)–(2.70). By construction, under the transformation Φ, one has the identity

$$\mathrm{tr}\left[V(L_N^\diamond)\right] \circ \Phi = \sum_{j=1}^{N} V(\lambda_j). \tag{9.28}$$

Inserting from (2.68), (2.69), in addition, one obtains

$$\mathrm{e}^{-q_1} \circ \Phi = \frac{\sigma_{N-1}}{\sigma_N} = \sum_{j=1}^{N} Y_j \mathrm{e}^{-\phi_j}, \quad \mathrm{e}^{q_N} \circ \Phi = \frac{\sigma_1}{\sigma_0} = \sum_{j=1}^{N} Y_j \mathrm{e}^{\phi_j} \qquad (9.29)$$

with

$$Y_j = \prod_{i=1, i \neq j}^{N} |\lambda_i - \lambda_j|^{-1}. \qquad (9.30)$$

We first consider the case when the average stretch $\nu > 0$. To study GGEs, the end particles are subject to suitable boundary potentials, $V_{\ell,1}^{+}(q_1)$, $V_{\ell,N}^{+}(q_N)$, such that all particles are confined to a volume of approximate size ℓ with $\ell > 0$. Our observation above suggests to choose the boundary potential as

$$V_{\ell,1}^{+}(q_1) + V_{\ell,N}^{+}(q_N) = \mathrm{e}^{-\ell/2}\left(\mathrm{e}^{-q_1} + \mathrm{e}^{q_N}\right), \qquad (9.31)$$

since e^{-q_1} and e^{q_N} are given by rather simple formulas. Then, in approximation, $-\ell/2 < q_1$ and $q_N < \ell/2$, which yields the desired confinement because $\nu > 0$. In case $\nu < 0$, one has to choose $\ell < 0$ and the boundary potential becomes

$$V_{\ell,1}^{-}(q_1) + V_{\ell,N}^{-}(q_N) = \mathrm{e}^{\ell/2}\left(\mathrm{e}^{q_1} + \mathrm{e}^{-q_N}\right). \qquad (9.32)$$

But now Y_j is switched to the denominator, which makes the analysis more delicate. In the following, we restrict ourselves to the case $\nu > 0$.

Considered next is a canonical-type ensemble with ℓ and N as parameters. Since Φ is a canonical transformation, its Jacobian determinant equals 1 which implies for the volume elements $\mathrm{d}^N p\, \mathrm{d}^N q = \mathrm{d}^N \phi\, \mathrm{d}^N \lambda$. To the energy-like term $\mathrm{tr}[V(L_N^\diamond)]$, one adds the boundary potential (9.31). For the partition function, one thereby arrives at

$$Z_{\mathrm{to},N}(\ell, V) = \int_{\Gamma_N} \mathrm{d}^N p\, \mathrm{d}^N q \exp\left(-\mathrm{tr}[V(L_N^\diamond)] - \mathrm{e}^{-\ell/2}\left(\mathrm{e}^{-q_1} + \mathrm{e}^{q_N}\right)\right)$$

$$= \frac{1}{N!} \int_{\mathbb{R}^N} \mathrm{d}^N \phi \int_{\mathbb{R}^N} \mathrm{d}^N \lambda \exp\left(-\sum_{j=1}^{N} V(\lambda_j) - 2\mathrm{e}^{-\ell/2} Y_j \cosh \phi_j\right).$$

$$(9.33)$$

Here, we used that under the map Φ^{-1}, the eigenvalues are ordered. Since the integrand is symmetric in the action variables, their integration is extended over $\mathbb{R}^N$ at the expense of the factor $1/N!$. The ϕ_j-integrations can be carried out explicitly and yield

$$Z_{\text{to},N}(\ell,V) = \frac{1}{N!} \int_{\mathbb{R}^N} \mathrm{d}^N \lambda \prod_{j=1}^N e^{-V(\lambda_j)} \prod_{j=1}^N 2K_0\big(2e^{-\ell/2}Y_j\big), \qquad (9.34)$$

where

$$K_0(x) = \int_0^\infty \mathrm{d}t\, \exp(-x\cosh t) \qquad (9.35)$$

is the zero order modified Bessel function of the second kind.

As a consequence, the probability distribution of eigenvalues of the Lax matrix $L_N^\diamond$ under the normalized measure

$$\frac{1}{Z_{\text{to},N}(\ell,V)} \exp\big(-\operatorname{tr}\big[V(L_N^\diamond)\big] - e^{-\ell/2}\big(e^{-q_1} + e^{q_N}\big)\big)\mathrm{d}^N p\,\mathrm{d}^N q \qquad (9.36)$$

is given by

$$\frac{1}{Z_{\text{to},N}(\ell,V)} \exp\left[-\sum_{j=1}^N V(\lambda_j)\right]$$

$$\times \exp\left[\sum_{j=1}^N \log\left(2K_0\left(2\exp\left[-\frac{1}{2}N\left(\nu - \frac{2}{N}\sum_{i=1,i\neq j}^N \log Y_j\right)\right]\right)\right)\right] \qquad (9.37)$$

relative to $(1/N!)\mathrm{d}^N\lambda$. Note that

$$-\frac{2}{N}\log Y_j = \frac{1}{N}\sum_{i=1,i\neq j}^N \phi_{\text{to}}(\lambda_j - \lambda_i) \qquad (9.38)$$

which, except for the diagonal term, equals $\phi_{\text{to}}(\lambda_j - \cdot)$ integrated over the empirical DOS. Only the functional in (9.37) is more complicated than before and has an explicit N-dependence in addition.

The next step is the $\ell \to \infty$ asymptotics at fixed $\nu = \ell/N$. The Bessel function behaves as $K_0(x) = -\log x$ for $x \to 0$ and $K_0(x) = (\pi/2x)^{1/2}e^{-x}$ for $x \to \infty$. Hence, for large ℓ,

$$K_0(e^{-a\ell}) \simeq a\ell \ \text{ for } a > 0, \quad \log K_0(e^{-a\ell}) \simeq -e^{-a\ell} \ \text{ for } a < 0. \qquad (9.39)$$

Let us define

$$\Upsilon_j = \nu + \frac{1}{N}\sum_{i=1,i\neq j}^N \phi_{\text{to}}(\lambda_j - \lambda_i). \qquad (9.40)$$

Then, for large N,

$$\log\left(2K_0\left(2\exp\left[-\frac{1}{2}N\Upsilon_j\right]\right)\right) \simeq \begin{cases} \log N + \log\Upsilon_j, & \text{if } \Upsilon_j > 0, \\ -2e^{-(N\Upsilon_j/2)}, & \text{if } \Upsilon_j < 0. \end{cases} \tag{9.41}$$

Following the arguments in Section 3.5, the corresponding free energy functional reads

$$\mathcal{F}^\circ_{\text{to,f},+}(\varrho) = \nu^{-1}\int_{\mathbb{R}} dw \varrho(w)\left(V(w) - 1 + \log\varrho(w)\right.$$

$$\left. - \log\left(\nu + \int_{\mathbb{R}} dw' \varrho(w')\phi_{\text{to}}(w - w')\right)\right) \tag{9.42}$$

in case $\nu + \int_{\mathbb{R}} dw' \varrho(w')\phi_{\text{to}}(w - w') > 0$ and $\mathcal{F}^\circ_{\text{to,f},+}(\varrho) = \infty$ otherwise. The variation is over all probability densities with $\mathcal{F}^\circ_{\text{to,f},+}(\varrho) < \infty$. Momentarily, we ignore this infinite-dimensional constraint but will return to the issue in the following. Denoting the minimizer by ϱ^*, as before, one concludes that

$$\lim_{\ell\to\infty} -\frac{1}{\ell}\log Z_{\text{to},N}(\ell, V) = \mathcal{F}_{\text{to,f},+}(\nu, V) = \mathcal{F}^\circ_{\text{to,f},+}(\varrho^*). \tag{9.43}$$

It is convenient to substitute $\nu^{-1}\varrho = \rho$. Then, the free energy per unit length of the Toda fluid is determined by the free energy functional

$$\mathcal{F}_{\text{to,f},+}(\rho) = \int_{\mathbb{R}} dw \rho(w)\left(V(w) - 1 + \log\rho(w)\right.$$

$$\left. - \log\left(1 + \int_{\mathbb{R}} dw' \rho(w')\phi_{\text{to}}(w - w')\right)\right), \tag{9.44}$$

where the minimization is constrained by $\rho \geqslant 0$ and

$$\int_{\mathbb{R}} dw \rho(w) = \nu^{-1}. \tag{9.45}$$

As before, the constraint is lifted by the Lagrange multiplier μ. Introducing the *space density* ρ_{s} by

$$\rho_{\text{s}} = 1 + T\rho, \tag{9.46}$$

the free energy functional can be written as

$$\mathcal{F}^\bullet_{\text{to,f},+}(\rho) = \int_{\mathbb{R}} dw \rho(w)\left(V(w) - 1 - \mu\right) + \int_{\mathbb{R}} dw \rho_{\text{s}}(w)\left(\frac{\rho}{\rho_{\text{s}}}\log\frac{\rho}{\rho_{\text{s}}}\right)(w). \tag{9.47}$$

The first summand can be viewed as energy-like and the second summand as the entropy of ρ relative to ρ_{s}.

Provisionally, we denote the minimizer of (9.47) by ρ_p which satisfies the Euler–Lagrange equation

$$V(w) - \mu + \log \frac{\rho_p}{1 + T\rho_p} - T\frac{\rho_p}{1 + T\rho_p} = 0. \tag{9.48}$$

In view of (3.61) and (3.49), this suggests to set

$$\rho_n = \frac{\rho_p}{1 + T\rho_p}, \quad \rho_n = e^{-\varepsilon}, \tag{9.49}$$

which leads to

$$\varepsilon = V - \mu - Te^{-\varepsilon}, \tag{9.50}$$

in agreement with the TBA equation (3.56). Since $\rho_s = 1 + T\rho_p$, the space density satisfies

$$\rho_s = \varsigma_0^{dr}, \quad \rho_s\rho_n = \rho_p. \tag{9.51}$$

In conclusion,

$$\lim_{\ell \to \infty} -\ell^{-1}\log Z_{to,f,+,N}(\ell, \mu, V) = \mathcal{F}_{to,f,+}^{\bullet}(\rho_p) = -\int_{\mathbb{R}} dw\, e^{-\varepsilon(w)}, \tag{9.52}$$

in agreement with (9.11).

As explained in Section 3.5, the convergence of the DOS to a deterministic limit follows on general grounds but can also be confirmed by considering the first-order variational derivatives of the free energy. Returning to the infinite-dimensional constraint below Eq. (9.42), evaluating it at ρ_p reads

$$0 < 1 + \int_{\mathbb{R}} dw'\rho_p(w')\phi_{to}(w - w') = \rho_s(w). \tag{9.53}$$

For $P < P_c$, the number density $\rho_n > 0$ and also ρ_p, ρ_s stay positive. Hence, (9.53) is equivalent to $\nu > 0$, a condition which was required already at the very beginning of the argument. Thus, indeed, the minimizer remains unaffected by the constraint.

The free energy for the Toda fluid has also been obtained by using scattering coordinates in combination with a specifically designed external potential. We presented two very different proofs for the generalized free energy and arrived at the same answer, as it should be. Since the proofs are lengthy, one might worry about missing factors of 2 or so. Having two independent derivations vouches for correctness. Furthermore, the availability of a second tool opens up the possibility to handle the generalized free energy for a larger class of integrable models.

◆◆ *Local macroscopic conservation laws:* By locally conserved we mean here the identity

$$\partial_t \mathfrak{n}(x, t; w) + \partial_x \mathfrak{j}(x, t; w) = 0 \tag{9.54}$$

pointwise in w. If GHD is considered in a periodic interval, say $[0, \ell]$, this implies the conservation law

$$\partial_t \int_0^\ell \mathrm{d}x\, \mathfrak{n}(x, t; w) = 0. \tag{9.55}$$

According to (9.47), a physical example is the entropy with density

$$s(x, t; w) = -\rho_{\mathsf{p}}(x, t; w) \log \rho_{\mathsf{n}}(x, t; w). \tag{9.56}$$

If the solution to GHD is free of discontinuities, the entropy is locally conserved.

Surprisingly, as a general property of GHD, beyond entropy, a much larger family of local macroscopic conservation laws is obtained in replacing the log by some arbitrary function, g, hence

$$\mathfrak{n}(x, t; w) = (\rho_{\mathsf{p}} g(\rho_{\mathsf{n}}))(x, t; w). \tag{9.57}$$

To verify, one starts from

$$-\partial_t(\rho_{\mathsf{p}} g(\rho_{\mathsf{n}})) = -(\partial_t \rho_{\mathsf{p}}) g(\rho_{\mathsf{n}}) - \rho_{\mathsf{p}} g'(\rho_{\mathsf{n}}) \partial_t \rho_{\mathsf{n}}$$
$$= \partial_x(v^{\mathrm{eff}} \rho_{\mathsf{p}}) g(\rho_{\mathsf{n}}) + \rho_{\mathsf{p}} v^{\mathrm{eff}} \partial_x g(\rho_{\mathsf{n}}) = \partial_x(v^{\mathrm{eff}} \rho_{\mathsf{p}} g(\rho_{\mathsf{n}})), \tag{9.58}$$

which implies

$$\mathfrak{j} = v^{\mathrm{eff}} \rho_{\mathsf{p}} g(\rho_{\mathsf{n}}), \tag{9.59}$$

as to be confirmed.   ◆◆

Notes and references

Section 9.1

A more detailed discussion of the Toda fluid is presented by Doyon (2019, 2020). The high-low pressure domain wall is studied by Mendl and Spohn (2022).

Section 9.2

The Toda lattice with boundary potential $g_1(\mathrm{e}^{-q_1} + \mathrm{e}^{q_N}) + g_2(\mathrm{e}^{-2q_1} + \mathrm{e}^{2q_N})$ is still integrable, see van Diejen (1995) for a proof. But the respective Lax matrix is less accessible. Using scattering coordinates for the computation of the generalized free energy is anticipated in Doyon (2019), where the free energy functional (9.42) can be found already. The identity (9.33) seems to be a novel result. The large class of local conservation laws for the hydrodynamic equations has been noted by Caux *et al.* (2019). I learned about (9.58) through the master thesis of Deokule (2022).

Chapter 10

Hydrodynamics of Soliton Gases

In his seminal contribution, Morikazu Toda realized that a chain with exponential interactions allows for traveling waves, called *soliton* in analogy to continuum nonlinear wave equations. The soliton is a coherent motion of many particles which is spatially localized and travels at constant velocity maintaining its shape. Two incoming solitons undergo an intricate collision process, to eventually move outward with velocities unchanged and characteristic scattering shifts. But qualitatively, this resembles the dynamics discussed for Toda quasiparticles already. Apparently, the Toda dynamics possesses two distinct levels with soliton-like structures. To distinguish them, we refer to either soliton-based schemes or particle-based schemes. The much studied prime example for a soliton-based scheme is the Korteweg–de Vries (KdV) equation. This topic will be taken up first because of its similarity with the soliton-based scheme of the Toda dynamics. Allowing for both hydrodynamic schemes is shared by Toda and Ablowitz–Ladik lattice. For other models, only either one of the two schemes is known.

10.1 Soliton gas of the KdV equation

The most widely studied soliton gas is based on the KdV equation, written in its conventional form as

$$\partial_t u + \partial_x (3u^2 + \partial_x^2 u) = 0 \tag{10.1}$$

with u a real-valued wave field over $\mathbb{R}$. Physically, the KdV equation describes the motion of surface waves in a shallow channel of water under gravity and in the approximation of small amplitudes. Solitary waves have been observed experimentally back in 1834 and the description in terms of a specific nonlinear wave equation was accomplished in 1895. While the study of soliton solutions strongly indicated integrability, the infinite hierarchy of local conservation laws was first obtained by Miura, Gardner, and Kruskal in 1968.

As the Toda lattice, KdV admits for a Lax pair. In fact, L is a Schrödinger operator acting on $L^2(\mathbb{R}, \mathrm{d}x)$,

$$L = -\partial_x^2 - u(x,t) \tag{10.2}$$

with $u(x,t)$ at fixed t understood as multiplication operator. The partner operator reads

$$B = -\left(4\partial_x^3 + 6u\partial_x + 4\partial_x u\right). \tag{10.3}$$

Under the KdV time evolution,

$$\partial_t L = [B, L], \tag{10.4}$$

a property signaling integrability. A different viewpoint is to consider, for a given u, the eigenvalue problem

$$Lv = \lambda v \tag{10.5}$$

and the evolution equation

$$\partial_t v = Bv. \tag{10.6}$$

This system has solutions only if $u(t)$ satisfies the KdV equation, which thus plays the role of a consistency relation.

The lowest order conserved fields can still be obtained by hand with the resulting densities

$$Q^{[0]}(x) = u(x), \quad Q^{[1]}(x) = u^2(x), \quad Q^{[2]}(x) = u^3(x) - \tfrac{1}{2}(\partial_x u)^2(x). \tag{10.7}$$

Here, we adopted that convention that u^{n+1} is the term with no derivatives. A more systematic tool comes from the underlying Poisson structure. The formal Poisson bracket turns out to be

$$\{u(x), u(x')\}_{\mathrm{KV}} = \delta'(x - x'). \tag{10.8}$$

More conveniently, functions on phase space are replaced by functionals of u and its derivatives,

$$F = \int_{\mathbb{R}} \mathrm{d}x h(u(x), \partial_x u(x), \ldots, \partial_x^n u(x)) \tag{10.9}$$

with some polynomial h. Of course, more general functionals could be considered. But for our purposes, (10.9) will do. The gradient of F, denoted by $\nabla F(x)$, is the functional derivative of F with respect to u at x. The functionals F, G have then the Poisson bracket

$$\{F, G\}_{\text{KV}} = \int_{\mathbb{R}} \mathrm{d}x \nabla F(x) \partial_x \nabla G(x). \tag{10.10}$$

In particular, this Poisson bracket satisfies the Jacobi identity. Denoting the total conserved fields by $Q^{[n]} = \int \mathrm{d}x Q^{[n]}(x)$, their respective gradients are

$$\nabla Q^{[0]} = 1, \quad \nabla Q^{[1]} = 2u, \quad \nabla Q^{[2]} = 3u^2 + \partial_x^2 u. \tag{10.11}$$

Higher-order conservation laws are obtained recursively through

$$(2n-1)\partial_x \nabla Q^{[n+1]} = (n+1)\left(u\partial_x + \partial_x u - \tfrac{1}{2}\partial_x^3\right)\nabla Q^{[n]}. \tag{10.12}$$

The prefactors result from our convention in (10.7) upon disregarding the operator ∂_x^3. In fact, this kind of recursion is familiar from integrable quantum many-body systems and is called *boost operator*.

For the Hamiltonian, we set $H_{\text{kv}} = -Q^{[2]}$, which then generates the time evolution as

$$\partial_t u(x, t) = \{u(x, t), H_{\text{kv}}\}_{\text{KV}}. \tag{10.13}$$

Hence, $Q^{[2]}$ is called energy. Analogously, $Q^{[1]}$ generates a spatial shift to the right with unit speed, hence called momentum. The KdV equation possesses a two-parameter family of traveling wave solutions, the *solitons*, more specifically one-solitons. Their analytic form is

$$u_{\text{s},1}(x, t; \eta, \delta) = 2\eta^2 \text{sech}^2\left(\eta(x - 4\eta^2 t - \delta)\right). \tag{10.14}$$

Here, η is the spectral parameter, $\eta > 0$, and δ is a shift, $\delta \in \mathbb{R}$. The center of the soliton moves on the straight line $x(t) = 4\eta^2 t + \delta$. Hence, the soliton speed is $4\eta^2$. The taller and narrower the shape, the larger the soliton speed. The conserved fields take the values $Q^{[0]}(u_{\text{s},1}) = 4\eta$, $Q^{[1]}(u_{\text{s},1}) = (4^2/3)\eta^3$, and $Q^{[2]}(u_{\text{s},1}) = (4^3/5)\eta^5$.

We now start with localized initial data, in the sense that

$$\int_{\mathbb{R}} \mathrm{d}x |x| |u(x)| < \infty. \tag{10.15}$$

For long times, $t \to \pm\infty$, the KdV solution splits into a soliton part, denoted by u^{s}, and a dispersive part which decays to zero with some power law. More precisely, for any $\epsilon > 0$,

$$\lim_{|t| \to \infty} \sup_{\{x \mid |x| > \epsilon |t|\}} |u(x, t) - u^{\text{s}}(x, t)| = 0. \tag{10.16}$$

This behavior is reminiscent of charges coupled to the Maxwell field which asymptotically acquire straight line motion through emitting radiation.

Hence, the dispersive part is also called radiation. The soliton part consists of N freely moving solitons, N depending on the initial condition. In a formula,

$$u^{\mathrm{s}}(x,t) = \sum_{j=1}^{N} u_{\mathrm{s},1}(x,t;\eta_j,\phi_j^{\pm}),\qquad(10.17)$$

where $\pm$ refers to $t \to \pm\infty$. Solitons behave like quasiparticles. However, for KdV, there are only overtaking collisions. Thus, for $t \to -\infty$, the incoming solitons are ordered as $x_j(t) < x_{j+1}(t)$ and the spectral parameters as $\eta_1 > \cdots > \eta_N > 0$, while outgoing solitons travel in reverse order. The label N is used in analogy to Section 2.3, where N denotes the number of quasiparticles (which equals the number of physical particles).

The crucial dynamical information comes from the scattering shifts, κ_j, which are written as

$$\phi_j^{-} \ \text{ for } t \to -\infty, \quad \phi_j^{+} = \phi_j^{-} + \kappa_j \ \text{ for } t \to \infty,\qquad(10.18)$$

where

$$\kappa_j = \frac{1}{\eta_j} \sum_{i=1,i\neq j}^{N} \mathrm{sgn}(\eta_j - \eta_i)\phi_{\mathrm{kv}}(\eta_j,\eta_i)\qquad(10.19)$$

with the KdV scattering kernel

$$\phi_{\mathrm{kv}}(\eta,\eta') = \log\left|\frac{\eta - \eta'}{\eta + \eta'}\right|.\qquad(10.20)$$

The latter structure is already familiar from the hard rod gas and Toda lattice. As a hallmark of integrability, the scattering shifts are determined by the two-soliton scattering shift. But in contrast to hard rods, the latter shift depends on the incoming velocities. Note that $\phi_{\mathrm{kv}}(\eta,\eta') < 0$ as for physical hard rods.

There are concise formulas for N-soliton solutions, which are defined by requiring the radiation part to vanish identically. Considering the special case $N = 2$, we can illustrate how the identity in (10.20) is obtained. The two-soliton solution can be written as

$$u_{\mathrm{s},2}(x,t) = 2\partial_x^2 \log \varphi_2(x,t)\qquad(10.21)$$

with

$$\varphi_2 = 1 + A_1 e^{2\theta_1} + A_2 e^{2\theta_2} + A_3 e^{2(\theta_1+\theta_2)},\qquad(10.22)$$

where

$$\theta_i = \eta_i x - \omega_i t, \quad \omega_i = 4\eta_i^3, \quad i = 1,2.\qquad(10.23)$$

The coefficients A_1, A_2 are free parameters which can be converted into $\phi_i^{\pm}$. But A_3 is determined by

$$\log A_3 = \log A_1 + \log A_2 + \phi_{\mathrm{kv}}(\eta_1, \eta_2). \tag{10.24}$$

In (10.19), the actual scattering shift carries the prefactor $1/\eta_j$. This dependence can be traced to the solution ansatz (10.23), where in the definition of θ_i, the factor η_i has not been bracketed out.

Given previous experience, one may boldly jump ahead to the hydrodynamic equation for the KdV soliton gas. We introduce the counting function $f(x,t;\eta)\mathrm{d}x\mathrm{d}\eta$ which is the number of solitons with spectral parameter in the interval $[\eta, \eta + \mathrm{d}\eta]$ and location in $[x, x + \mathrm{d}x]$ at time t. The counting function is governed by

$$\partial_t f + \partial_x(v^{\mathrm{eff}} f) = 0. \tag{10.25}$$

For the effective velocity, as before, we assume the collision rate ansatz which in our case becomes

$$v^{\mathrm{eff}}(\eta) = 4\eta^2 + \eta^{-1} \int_0^\infty \mathrm{d}\eta \phi_{\mathrm{kv}}(\eta, \eta') f(\eta')\big(v^{\mathrm{eff}}(\eta') - v^{\mathrm{eff}}(\eta)\big). \tag{10.26}$$

Still missing are some explanations on how the counting function f is related to local values of the conserved fields. In principle, the counting function can be measured by the time-of-flight method. One considers some time t and a sufficiently large spatial box, still small on the scale of macroscopic variations. The wave field $u(x)$ in this spacetime cell is cut out and then smoothly interpolated with the zero field all the way to infinity. Such a constructed field $u_{\mathrm{box}}(x)$ is evolved under the KdV dynamics. For sufficiently long times, the solution consists of many spatially well-separated one-solitons. In approximation, their statistics is that of an ideal gas with uniform density $\bar{\varrho} = \int_{\mathbb{R}_+} \mathrm{d}\eta f(\eta)$. Each soliton carries a spectral parameter, η_j, and these form a family of i.i.d. random variables with single point probability density $h(\eta) = f(\eta)/\bar{\varrho}$. To determine the average value of the conserved field $Q^{[n]}$ under such statistics, the positional average yields the integral over $Q^{[n]}$ evaluated at $u_{\mathrm{s},1}(x,0;\eta,0))$ and the i.i.d. average results in the integral over η weighted with $h(\eta)$, in formulas

$$\int_0^\infty \mathrm{d}\eta h(\eta)\bar{\varrho} \int_{\mathbb{R}} \mathrm{d}x Q^{[n]}(u_{\mathrm{s},1}(x,0;\eta,0)) = \frac{2^{2n+2}}{2n+1} \int_0^\infty \mathrm{d}\eta f(\eta)\eta^{2n+1}. \tag{10.27}$$

Thereby we connected the average of the nth conserved field with the $(2n+1)$th moment of the counting function. The numerical coefficients are known as Kruskal series. Since f vanishes on $\mathbb{R}_-$, it is uniquely determined by its moments (10.27).

A natural question is to have a more intrinsic characterization of the statistics of the wave field $u_{\mathrm{box}}(x)$ in the prescribed cell. This is so as to speak an inverse scattering problem and not much is known. For particle-based hydrodynamics, the statistics of the positions and momenta of the particles in the cell should be given by a GGE with properly adjusted parameters. At least for the Toda fluid we know that, with regard to the conserved fields, the data agree with those from the time-of-flight method. Also, the Lax filter could be employed, compare with Section 3.2. For KdV, the corresponding construction would involve bound state eigenvalues of the Schrödinger operator (10.2) with potential given by the box field $u_{\mathrm{box}}(x)$.

•• *TBA revisited*: The Toda lattice consists of nonrelativistic particles with energy–momentum relation $E(P) = P^2/2$. For relativistic models, the energy–momentum relation would have to be modified. Also, for solitons, there is no particular reason to have a quadratic energy–momentum relation. Therefore, TBA has to be extended to cover such cases.

In a Hamiltonian context, if the energy–momentum relation $E(P)$ is given, then the velocity is determined by $v = \partial E/\partial P$. In case of the KdV equation, the one-soliton momentum and energy are given in terms of the spectral parameter η. To stay in context, we still use η, which however is now a parameter depending on the model under study. The functions $E(\eta)$ and $P(\eta)$ are prescribed. Then, with prime referring to parameter differentiation, the velocity becomes $v = E'/P'$. Through the interaction between quasiparticles, E', P' become dressed and, appealing to the fundamental relation (6.20), one argues that

$$v^{\mathrm{eff}} = \frac{(E')^{\mathrm{dr}}}{(P')^{\mathrm{dr}}}, \tag{10.28}$$

the dressing transformation being defined as before. Of course, for $E(P) = P^2/2$, the original relation is recovered.

Starting from the collision rate ansatz, it is now reformulated as

$$v^{\mathrm{eff}} = \frac{1}{P'}\big(E' + T(\rho_{\mathsf{p}}v^{\mathrm{eff}}) - (T\rho_{\mathsf{p}})v^{\mathrm{eff}}\big), \tag{10.29}$$

where $T(\eta, \eta')$ is some symmetric kernel. As before, ρ_{p} is the particle density. The number density is defined by

$$\rho_{\mathsf{n}} = \frac{\rho_{\mathsf{p}}}{P' + T\rho_{\mathsf{p}}}. \tag{10.30}$$

Then, the dressing operation for some function f is still

$$f^{\mathrm{dr}} = (1 - T\rho_{\mathsf{n}})^{-1} f \qquad (10.31)$$

with ρ_{n} regarded as multiplication operator. The inverse to (10.30) reads

$$\rho_{\mathsf{p}} = \rho_{\mathsf{n}} P'^{\mathrm{dr}}, \qquad (10.32)$$

which identifies the space density as

$$\rho_{\mathsf{s}} = P'^{\mathrm{dr}}. \qquad (10.33)$$

By a computation similar to the one in (6.21), the effective velocity is indeed given by (10.28).

Applying our scheme to KdV solitons, one finds $P(\eta) = \eta^2/2$ and $E(\eta) = \eta^4$. They differ from the one-soliton momentum and energy.　♦♦

10.2 Soliton gas of the Toda lattice

We have already studied the hydrodynamic scale of the Toda lattice within the particle-based scheme. Since the Toda lattice has also solitons in the more conventional sense, one would expect to have also a hydrodynamic scale for the soliton-based scheme. This is the topic to be explored now. Of course, which scheme to use depends on the choice of initial conditions. For sure, there will be also initial conditions not covered by either scheme.

The dynamical evolution takes place on the infinitely extended lattice with equations of motion

$$\frac{\mathrm{d}^2}{\mathrm{d}t^2} r_j(t) = 2\mathrm{e}^{-r_j(t)} - \mathrm{e}^{-r_{j-1}(t)} - \mathrm{e}^{-r_{j+1}(t)}, \qquad (10.34)$$

where $j \in \mathbb{Z}$. We consider solutions for which $|p_j(t)|, |r_j(t)| \to 0$ as $j \to \pm\infty$. In Flaschka variables, the two-parameter family of one-soliton solutions is given by

$$a_j^2(t; \eta, \delta) - 1 = \omega^2 \mathrm{sech}^{-2}(|\eta|j - \omega t - \delta) \qquad (10.35)$$

and

$$p_j(t; \eta, \delta) = |\omega|\big(\tanh(|\eta|j - \omega t - \delta) - \tanh(|\eta|(j-1) - \omega t - \delta)\big) \qquad (10.36)$$

with

$$\omega(\eta) = \sinh \eta. \qquad (10.37)$$

Deviating from the previous convention, the soliton travels along the trajectory $j(t) = |\eta|^{-1}(\omega t + \delta)$. The spectral parameter is η, $\eta \in \mathbb{R}$, the shift $\delta \in \mathbb{R}$, and the soliton velocity

$$v_s = |\eta|^{-1} \omega. \tag{10.38}$$

For $\eta > 0$, the soliton is moving to the right and for $\eta < 0$, to the left. Note that the velocity is odd and monotone increasing in η, however with a jump of size 2 at the origin. Necessarily, the soliton speed is supersonic.

The construction of the two-soliton relies on the discrete analog of (10.21) and the same ansatz as in (10.22), using $\theta_i = |\eta_i| - \omega_i t$. This results in the Toda scattering kernel denoted by $\phi_{\mathrm{tos}}(\eta_1, \eta_2)$. One has to distinguish overtaking collisions, $\eta_1 \eta_2 > 0$, and head-on collisions, $\eta_1 \eta_2 < 0$. Explicitly,

$$\phi_{\mathrm{tos}}^+(\eta_1, \eta_2) = 2 \log \left| \frac{\sinh \frac{1}{2}(\eta_1 - \eta_2)}{\sinh \frac{1}{2}(\eta_1 + \eta_2)} \right| \tag{10.39}$$

for $\eta_1 \eta_2 > 0$ and

$$\phi_{\mathrm{tos}}^-(\eta, \eta') = 2 \log \left| \frac{\cosh \frac{1}{2}(\eta_1 - \eta_2)}{\cosh \frac{1}{2}(\eta_1 + \eta_2)} \right| \tag{10.40}$$

for $\eta_1 \eta_2 < 0$. Clearly, on their domain of definition, $\phi_{\mathrm{tos}}^+ < 0$ as hard core particles with $a > 0$ and $\phi_{\mathrm{tos}}^-(\eta, \eta') > 0$ as hard core particles with $a < 0$.

According to the collision rate ansatz, the effective velocity for Toda solitons is

$$v^{\mathrm{eff}}(\eta) = |\eta|^{-1} \left(\sinh \eta + \int_{\mathbb{R}} \mathrm{d}\eta \phi_{\mathrm{tos}}(\eta, \eta') f(\eta') \left(v^{\mathrm{eff}}(\eta') - v^{\mathrm{eff}}(\eta) \right) \right). \tag{10.41}$$

In terms of Eq. (10.28), one infers that

$$P(\eta) = \tfrac{1}{2}\mathrm{sgn}(\eta)\eta^2, \quad E(\eta) = \cosh \eta. \tag{10.42}$$

Without any changes, we can now follow the blueprint of the KdV equation. In particular, the soliton-based hydrodynamic equations are given by (10.25) upon substituting ϕ_{kv} by ϕ_{tos} in Eq. (10.26).

For the interpretation of the soliton counting function, we have to work out the analog of (10.27). As can be seen from (10.35), in the Flaschka variables, the soliton does not decay to 0. Therefore, one has to subtract the constant at infinity as

$$Q^{[n]\infty} = \mathrm{tr}[L^n - (L_\infty)^n], \tag{10.43}$$

$n \geqslant 1$, with $(L_\infty)_{i,j} = 1$ for $i = j \pm 1$ and 0 otherwise. We denote by K_n the conserved field $Q^{[n]\infty}$ evaluated at the one-soliton solution written in (10.35) and (10.36). These coefficients can be obtained from the inverse scattering transform (IST). The result is not as explicit as the one in (10.27) and reads

$$\sum_{n=1}^{\infty} \frac{1}{n} \left(\frac{z}{1+z^2} \right)^n K_n(\eta) = \sum_{n=1}^{\infty} \frac{2}{n} z^n \sinh(n\eta) \qquad (10.44)$$

for $\eta \geqslant 0$. To extend to negative η, one notes the symmetry $K_n(\eta) = (-1)^n K_n(-\eta)$, which follows from the random walk expansion of the right-hand side of (10.43). Thus, to the lowest order,

$$Q^{[1]\infty} = 2 \sinh \eta, \quad Q^{[2]\infty} = 2|\sinh(2\eta)|. \qquad (10.45)$$

Higher orders are determined recursively, at least in principle. But the basis functions are no longer polynomials as in the case of KdV.

10.3 Comparing soliton- and particle-based hydrodynamics

This seems to be a good moment for a broader perspective, for which purpose various integrable many-body systems are listed, some of them only to be discussed in later chapters. In Table 10.1, we list the model system, whether it is soliton- or particle-based, and the method used for deriving GHD. Round brackets indicate that the respective property is likely but more work is required.

Table 10.1. IST stands for inverse scattering transform, sc-co for scattering coordinates, and RMT for random matrix theory.

Model	Soliton-based	Particle-based	Method
Hard rods	No	Yes	Model specific
Box-ball	Yes	No	Model specific
KdV	Yes	No	IST
Toda (classical)	Yes	Yes	IST, sc-co, RMT
Calogero (classical)	(Yes)	Yes	sc-co
Continuum NLS	Yes	No	IST
Ablowitz–Ladik	(Yes)	Yes	IST, RMT
Quantum	No	Yes	Bethe ansatz

In the context of classical models, IST is the key tool for soliton-based hydrodynamics. Clearly, hard rods do not possess solitons. Box-ball is a spacetime discrete version of KdV. Configurations are two-sided infinite sequences of 0s and 1s. For a finite number of particles, the update from time t to $t+1$ is given by the following rule: an empty carrier with unbounded capacity starts from the very left, sequentially hops to the right, and at every site either picks up a particle or drops a particle whenever such move is possible. After the carrier has traveled to the very right, the time $t+1$ configuration is completed. An N-soliton is simply a stretch of N consecutive particles, isolated and with no holes. The soliton speed equals N. Soliton-based dynamics is confirmed both through mathematical proofs and numerical simulations. Most remarkably, the collision rate ansatz remains valid. Based on our discussions in Section 10.1, KdV is soliton based and so is the nonlinear Schrödinger equation (NLS).

Less clear-cut is the situation for the Calogero model with $1/\sinh^2$ interaction potential. Since the Toda lattice is a low-density approximation to the Calogero model, one might have expected at least the analytic form of one-solitons. This is not the case, but promising numerical results are recorded in the literature. For the Ablowitz–Ladik system, to be discussed in Chapter 12, N-soliton solutions are available through IST and also the two-soliton scattering shift. But apparently, soliton-based hydrodynamics has been left out. Presumably efforts have been focussed more strongly on the study of the hydrodynamic scale of the KdV and continuum nonlinear Schrödinger equation because of their relevance for experimental realizations.

For particle-based hydrodynamics, a prerequisite is the availability of a Lax matrix. But then two distinct methods are used. One is the transformation to scattering coordinates, as explained already for the Toda fluid. An alternative method emerges from the observation that under GGE the Lax matrix becomes a random matrix. Thereby tools from random matrix theory (RMT) become available. This method works well for the Toda lattice and also for the Ablowitz–Ladik model. In the latter case, the Lax matrix is unitary and the eigenvalues lie on the unit circle. In fact, the Ablowitz–Ladik model plays a peculiar role. From the perspective of RMT it is obviously particle-based. But in contrast to the Toda lattice and Calogero fluid, the true particle trajectories are hidden.

In deriving GHD for quantum many-body systems, the very first step toward an analysis is always the Bethe ansatz, which determines the eigenvalues of the many-body system, compare with Chapter 13. Thereby generalized thermodynamics can be constructed. Since in the respective Bethe equations the two-particle phase shift appears explicitly, it is not so surprising that this scattering shift reappears in the generalized free energy. Here, two-particle phase shift refers to the dynamics in the two-particle subspace. In our models, this would be the Hilbert space $L^2(\mathbb{R}^2, \mathrm{d}x_1 \mathrm{d}x_2)$ and the respective quantum wave functions are of the form $\psi(x_1, x_2, t)$. Thus, clearly, for integrable quantum many-body systems GHD is particle-based.

In analogy to IST, the Faddeev school developed the quantum inverse scattering method (QISM). This is a very powerful tool from which a unifying view of integrability emerged, encompassing also integrable models from two-dimensional statistical mechanics. Apparently, the method was never used to construct a soliton gas, say like the one for the KdV equation. There is interesting work around the notion of quantum solitons. Whether these objects are really solitons in the sense of scattering theory remains to be explored.

Notes and references

Section 10.1

The integrability of the KdV equation has been established in the late 1960s in a series of contributions by Gardner *et al.* (1967), Miura *et al.* (1968), and Kruskal *et al.* (1970). The Hamiltonian structure is discussed in Gardner (1971). An important step was the development of inverse scattering transform, which was then applied to a variety of other models. We refer to the review by Ablowitz *et al.* (1974) and the books by Ablowitz and Segur (1981) and Ablowitz *et al.* (2004). Scattering theory of the KdV equation is studied by Tanaka (1975, 1976). Whitham (1974) developed his modulation theory in the context of KdV, see also Flaschka *et al.* (1980). Soliton gases were first introduced by Zakharov (1971), see also Zakharov (2009). He used a low-density approximation, for which in the collision rate ansatz (10.26), the term v^{eff} on the right-hand side is replaced by its bare value $4\eta^2$. The hydrodynamic scale was pioneered by Gennady El in his seminal contributions El (2003) and El and Kamchatnov (2005). El uses the Whitham modulation theory in a limit where the spectral bands of $L(t)$ become very flat, see El (2021) for a very instructive review. In fact, the collision rate ansatz appears only indirectly through the coupled equations

$$-Tf(\eta) + \sigma(\eta)f(\eta) = \eta, \quad -Tv(\eta) + \sigma(\eta)f(\eta) = 4\eta^3, \qquad (10.46)$$

where $v = v^{\text{eff}} f$ is the spectral flux density. Eliminating σ yields Eq. (10.26) for v^{eff}. A further confirmation comes from numerical studies of a tracer soliton and from matching with hydrodynamic reductions, see Carbone *et al.* (2016). The latter means that the initial $f(x; \eta)$ is piecewise constant and takes only two or three values. Then, the Euler equations simplify to make explicit solutions available, which can be compared with numerical solutions of KdV. The initial conditions have spatial randomness but very little noise in the η dependence. A further, this time mathematical confirmation, is accomplished by Girotti *et al.* (2021). The initial state is a long periodic arrangement of one-solitons, in other words, the cnoidal wave with spatial cut-off. In addition, a tracer soliton is incoming from the left with a sufficiently high velocity when compared to the building blocks of the cnoidal wave. The initial data are not random. Still in simulations, the tracer motion fluctuates random-like relative to a constant drift. The outgoing tracer velocity is given by (10.26) with f referring to the cnoidal background.

A more detailed comparison between GHD and KdV soliton-based hydrodynamics can be found in Bonnemain *et al.* (2022). Bidirectional solitons are studied by Congy *et al.* (2021) and soliton condensates by Congy *et al.* (2023). Experimental realizations are reported by Redor *et al.* (2019).

Section 10.2

In part II of his famous papers on the Toda lattice, Flaschka (1974a) developed the inverse scattering transform for the Toda lattice. Hirota (1973) constructed the N-soliton solution. A very readable account is the article by Toda (1983), where the derivation of scattering shift is explained and the similarity with KdV outlined. Scattering theory is studied by Krüger and Teschl (2009). For the infinite lattice, one considers initial conditions such that

$$\sum_{j \in \mathbb{Z}} (1 + |j|)\big(|a_j - 1| + |p_j|\big) < \infty. \tag{10.47}$$

In the long time limit, both future and past, the solution converges to N freely propagating one-solitons with scattering shifts determined by (10.39), (10.40). There seems to be no concrete algorithm to determine N for the respective soliton velocities for a given initial configuration.

Section 10.3

The clear distinction between particle-based and soliton-based hydrodynamics seems to be a novel insight. In principle, the N-soliton solutions of the Calogero fluid can be constructed through a Bäcklund transformation, also known as dual particles, see Philip (2019). Numerical solutions convincingly show the motion of a single soliton in a trap and collision between two solitons, as reported by Gon and Kulkarni (2019). Named as "soliton cellular automaton" an ultra-discrete wave equation was introduced by Takahashi and Satsuma (1990). By now box-ball is the universally accepted label referring to the picture that the carrier picks up balls from filled boxes, resp. drops balls to empty boxes. The hydrodynamic scale of the box-ball system has been studied in considerable detail by Ferrari *et al.* (2021), Croydon and Sasada (2021), and Kuniba *et al.* (2020, 2021).

Korepin *et al.* (1997) is the standard monograph on QISM. Faddeev (1995) is an illuminating account on the history. His 1996 Les Houches lectures, Faddeev (1996), still stand out as a masterful introduction.

Chapter 11

Calogero Models

Based on the experience with the Toda lattice, a physically natural goal is to advance toward one-dimensional classical fluids. The standard Hamiltonian reads

$$H_{\mathrm{mec},N} = \sum_{j=1}^{N} \tfrac{1}{2} p_j^2 + \tfrac{1}{2} \sum_{i,j=1, i \neq j}^{N} V_{\mathrm{mec}}(q_i - q_j), \qquad (11.1)$$

using units such that the particle mass equals 1. The particles move on the real line, with possibly some confining mechanism to be introduced later. The interaction potential, V_{mec}, is even and hence the Hamiltonian is invariant under relabeling of particles. A proper hydrodynamic scale assumes a short range potential, say with exponential decay. The borderline for a liquid–gas phase transition would be $V_{\mathrm{mec}}(x) \approx -|x|^{-2}$ for large $|x|$. But to our knowledge, the minimal decay for hydrodynamic behavior has not been investigated systematically. At least, the free energy has to be well defined which essentially is ensured by the stability property

$$\sum_{i \neq j=1}^{N} V_{\mathrm{mec}}(q_i - q_j) \geqslant -\frac{c_0}{N} \qquad (11.2)$$

with some constant $c_0 \geqslant 0$ independent of N. For a purely attractive potential, such a condition is violated and typically particles would clump together.

For a generic choice of V_{mec}, one expects number, momentum, and energy to be the only locally conserved fields. As a consequence, the conserved fields are expected to be governed by the conventional hydrodynamic equations, which usually are written as

$$\partial_t \rho_{\mathsf{f}} + \partial_x (\rho_{\mathsf{f}} v) = 0,$$

$$\partial_t (\rho_{\mathsf{f}} v) + \partial_x (\rho_{\mathsf{f}} v^2 + P_{\mathsf{f}}) = 0,$$

$$\partial_t (\rho_{\mathsf{f}} \mathfrak{e}_{\mathrm{tot}}) + \partial_x (\rho_{\mathsf{f}} \mathfrak{e}_{\mathrm{tot}} v + P_{\mathsf{f}} v) = 0. \tag{11.3}$$

Here, ρ_{f} is the particle density per unit length, $\rho_{\mathsf{f}} v$ the momentum density, and $\rho_{\mathsf{f}} \mathfrak{e}_{\mathrm{tot}}$ the total energy density. P_{f} is the thermodynamic pressure that depends on ρ_{f} and the internal energy, the latter being obtained from $\mathfrak{e}_{\mathrm{tot}} = \mathfrak{e}_{\mathrm{int}} + \frac{1}{2} v^2$. These equations are similar to Euler type equations of a nonintegrable chain. Only, since the potential is a sum over all pairs of particles, it is considerably more difficult to compute the pressure P_{f}.

A priori, there might be no choice of V_{mec} which makes the model integrable. We discussed already a fluid with hard core potential, which is integrable but relies on having only contact interactions. In 1975, Francesco Calogero investigated the issue and discovered exactly two interaction potentials for which the N-particle dynamics is integrable: the long-ranged $1/x^2$ potential, called rational Calogero–Moser model, and the short-range repulsive $1/\sinh^2(x)$ potential, which we will refer to as Calogero fluid, resp. hyperbolic Calogero model.

◆◆ *Two distinct Lax matrices*: For a Hamiltonian as in (11.1), to check the integrability, Calogero assumes the existence of a Lax pair, L, M, of the general form

$$L_{i,j} = \delta_{ij} p_j + (1 - \delta_{ij}) \alpha(q_i - q_j),$$

$$M_{i,j} = \delta_{ij} \left(\sum_{k=1, k \neq j}^{N} \beta(q_j - q_k) \right) + (1 - \delta_{ij}) \gamma(q_i - q_j) \tag{11.4}$$

with yet arbitrary functions α, β, γ. The commutator condition for the Lax pair then yields functional relations for α, β, γ and the potential V_{mec}. These equations can be analyzed with the result $\gamma = \alpha'$, $\beta = \alpha''/(2\alpha)$, and $V_{\mathrm{mec}}(x) = \alpha(x)\alpha(-x)$. Now, for α, there are essentially only three choices: (i) $\alpha(x) = ig/x$, (ii) $\alpha(x) = iga/\sinh(ax)$, and (iii) $\alpha(x) = iga \coth(ax)$. Here, a, g are real parameters. Case (i) is the rational Calogero model, which is discussed in Section 11.6, while both (ii) and (iii) yield the dynamics governed by the hyperbolic Calogero model in the following section. In addition, there are solutions corresponding to Lax pairs which result from periodizing over a ring of given size. This then generates the interaction potential $1/\sin x$ in case (i) and particular elliptic functions in cases (ii) and (iii).

It is surprising that the Calogero fluid has two distinct Lax matrices, say L and $\tilde{L}$. Depending on the problem, either one of them might be of advantage. There seems to be no simple relation between L and $\tilde{L}$. But since their eigenvalues are conserved when evolved by the same dynamics, there must be a functional relation between either set of eigenvalues.

For chains, one can carry out a similar analysis. Since a smooth potential is assumed, only the exponential potential qualifies as being integrable. In fact, there are also two distinct Lax matrices, but in this case, they are related by a simple similarity transform. ◆◆

11.1 Hyperbolic Calogero model, charges and currents

The interaction potential of the Calogero fluid is given by

$$V_{\mathrm{ca}}(x) = \frac{1}{(2\sinh(\frac{1}{2}x))^2}. \tag{11.5}$$

More commonly, one introduces length and time scales through

$$V_{\mathrm{ca}}(x) = \frac{g^2 a^2}{\sinh^2(ax)}, \tag{11.6}$$

which would make the formulas more lengthy. If needed, these parameters can be introduced by a linear change of spacetime coordinates. The choices $g = 1, a = \frac{1}{2}$ are dictated by recovering the Toda lattice at low density. For N particles on the entire real line, Newton equations of motion are

$$\frac{\mathrm{d}^2}{\mathrm{d}t^2}q_j = \sum_{i=1,i\neq j}^{N} \tfrac{1}{4}\cosh(\tfrac{1}{2}(q_j - q_i))\big(\sinh(\tfrac{1}{2}(q_j - q_i))\big)^{-3} \tag{11.7}$$

for $j = 1, \ldots, N$. At short distances, the potential has the repulsive $1/|x|^2$ singularity, which implies that the spatial ordering of particles is maintained throughout time, compare with Figure 11.1.

The dynamics admits a Lax pair, L, M, through

$$L_{i,j} = \delta_{ij} p_j + \mathrm{i}(1 - \delta_{ij})\big(2\sinh(\tfrac{1}{2}(q_i - q_j))\big)^{-1},$$

$$M_{i,j} = \mathrm{i}\delta_{ij}\left(\sum_{k=1,k\neq j}^{N} \big(2\sinh(\tfrac{1}{2}(q_j - q_k))\big)^{-2}\right)$$

$$- \mathrm{i}(1 - \delta_{ij})\cosh(\tfrac{1}{2}(q_i - q_j))\big(2\sinh(\tfrac{1}{2}(q_i - q_j))\big)^{-2} \tag{11.8}$$

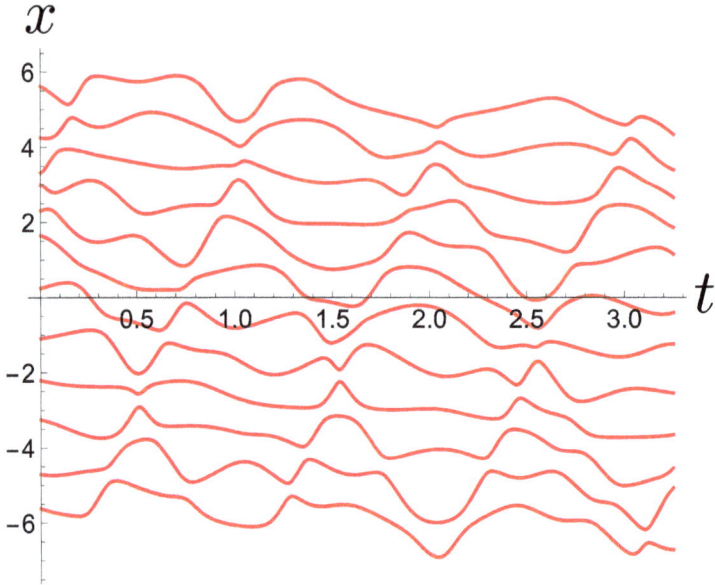

Fig. 11.1. Displayed are trajectories of 12 Calogero particles in a periodic box of size $\ell = 12$ and at temperature $\beta = 1$. Particles repel each other and cannot cross. There is one fast quasiparticle which can be distinguished from the background. The trajectories shown on the book cover are obtained from solving the equations of motion for various initial conditions. Courtesy from M. Kulkarni.

for $i, j = 1, \ldots, N$. Referring to the Insert above, the choice $\tilde{L}$ would lead to non-local conservation laws, hence not acceptable for hydrodynamics. For a lighter notation, we omit the index N that was used for the Toda lattice. The Lax matrix is Hermitian, hence has real eigenvalues, while the partner matrix is anti-Hermitian. Then, with L, M evaluated along trajectories of (11.7),

$$\frac{\mathrm{d}}{\mathrm{d}t} L(t) = [L(t), M(t)]. \tag{11.9}$$

Merely having a commutator implies that

$$\frac{\mathrm{d}}{\mathrm{d}t} \mathrm{tr}[L(t)^n] = 0. \tag{11.10}$$

The eigenvalues of $L(t)$ do not change in time. The conserved charges are

$$Q^{[n]} = \mathrm{tr}[L^n] \tag{11.11}$$

and their density is given by

$$Q^{[n]}(x) = \sum_{j=1}^{N} \delta(x - q_j)(L^n)_{j,j} \tag{11.12}$$

with $x \in \mathbb{R}$. The total momentum corresponds to $Q^{[1]}$ and total energy to $\frac{1}{2}Q^{[2]}$. As a general fact for Calogero-type models, under the standard Poisson bracket,

$$\{Q^{[m]}, Q^{[n]}\} = 0. \tag{11.13}$$

Thus, we have obtained a set of N conserved fields in involution. Setting $n = 0$ our labeling includes the particle number,

$$\text{tr}[L^0] = N = Q^{[0]}, \quad Q^{[0]}(x) = \sum_{j=1}^{N} \delta(x - q_j). \tag{11.14}$$

In the sense of Hamiltonian systems, the constant function is a trivial conservation law. But it must be included in a hydrodynamic context, since the particle density is a central physical observable.

To obtain the currents, one considers

$$\frac{d}{dt}Q^{[n]}(f) = \frac{d}{dt}\sum_{j=1}^{N} f(q_j)(L^n)_{j,j}$$

$$= \sum_{j=1}^{N}\left(f'(q_j)p_j(L^n)_{j,j} + \sum_{i=1,i\neq j}^{N}(f(q_j) - f(q_i))M_{i,j}(L^n)_{j,i}\right), \tag{11.15}$$

which implies

$$J^{[n]}(x) = \sum_{j=1}^{N}\left(\delta(q_j - x)p_j(L^n)_{j,j}\right.$$

$$\left. + \sum_{i=1,i\neq j}^{N}(\theta(q_j - x) - \theta(q_i - x))M_{i,j}(L^n)_{j,i}\right). \tag{11.16}$$

In particular,

$$\partial_t Q^{[n]}(x,t) + \partial_x J^{[n]}(x,t) = 0. \tag{11.17}$$

Most naturally the system is confined by having particles move in a periodic box $[0, \ell]$, $\ell > 0$. Since the interaction potential has a tail, the potential from all image particles has to be included. To introduce a notation, for some rapidly decreasing function, f, we define its ℓ-periodized version through

$$f_{\mathrm{per},\ell}(x) = \sum_{m \in \mathbb{Z}} f(x + m\ell). \tag{11.18}$$

By construction, $f_{\mathrm{per},\ell}$ is ℓ-periodic and its domain can be restricted to the box $[0, \ell]$. The periodized interaction potential can be written in terms of the elliptic (double periodic) Weierstrass function, $\wp(z)$. Historically, its two periodicities are denoted by $2\omega_1, 2\omega_2$ and

$$\wp(z|\omega_1, \omega_2) = \frac{1}{z^2} + {\sum_{m,n \in \mathbb{Z}}}' \left(\left(z + 2\omega_1 m + 2\omega_2 n\right)^{-2} - \left(2\omega_1 m + 2\omega_2 n\right)^{-2} \right),$$
$$\tag{11.19}$$

where $\sum'$ means that the origin, $(m, n) = 0$, has to be omitted in the sum. One notes the derivative

$$\wp(z|\omega_1, \omega_2)' = - \sum_{m,n \in \mathbb{Z}} 2\left(z + 2\omega_1 m + 2\omega_2 n\right)^{-3} \tag{11.20}$$

and the identities

$$\sum_{n \in \mathbb{Z}} \left(z + \mathrm{i}\pi n\right)^{-2} = \sinh^{-2}(z), \quad -\sum_{n \in \mathbb{Z}} 2\left(z + \mathrm{i}\pi n\right)^{-3} = (\sinh^{-2}(z))'.$$
$$\tag{11.21}$$

We now set $2\omega_1 = \ell$ and $2\omega_2 = \mathrm{i}\pi$ and conclude from (11.20) and (11.21) that

$$(\sinh^{-2})_{\mathrm{per},\ell}(x) = \wp(x|\tfrac{1}{2}\ell, \mathrm{i}\tfrac{1}{2}\pi) - \tfrac{1}{12}. \tag{11.22}$$

The constant $-\frac{1}{12}$ has to be fixed by a separate argument and drops out when studying the dynamics.

The equations of motion for the Calogero particles in a periodic box are still given by Eq. (11.7) with the force replaced by the periodized force. However, the Lax matrix for the ring is no longer simply the periodized Lax matrix. By the method indicated in the Insert above, one can still construct a Lax pair for the periodized Calogero model in terms of particular elliptic functions. As for the Toda chain, the Calogero model remains integrable on the ring.

•• *Dilute limit*: If particles are far apart, the interaction potential can be approximated as $\mathrm{e}^{-|x|}$. Labeling particles in increasing order, the force on the jth particle is exponentially dominated by the force from its two neighbors. Thus, in approximation,

$$\frac{\mathrm{d}^2}{\mathrm{d}t^2}q_j = \mathrm{e}^{-(q_j-q_{j-1})} - \mathrm{e}^{-(q_{j+1}-q_j)}, \tag{11.23}$$

which is recognized as the Toda dynamics.

One would expect that in the dilute limit also the Lax pair converges to the Lax matrix of the Toda lattice. In this approximation,

$$L_{j,j} = p_j, \quad L_{j,j+1} = -\mathrm{i}a_j, \quad L_{j+1,j} = \mathrm{i}a_j,$$

$$M_{j,j} = 0, \quad M_{j,j+1} = M_{j+1,j} = -\tfrac{1}{2}\mathrm{i}a_j, \tag{11.24}$$

where, so as to recall, the Flaschka variables are $a_j = \exp[-(q_{j+1} - q_j)/2]$. For the open system, all other matrix elements vanish, while for the closed system, $L_{1,N} = \mathrm{i}a_N$ and $L_{N,1} = -\mathrm{i}a_N$. One checks that for the closed system, there is a diagonal unitary matrix, $\tilde{U}$, such that $\tilde{U}L\tilde{U}^* = L_N$ and and $\tilde{U}M\tilde{U}^* = -B_N$, while for the open system, there is a diagonal unitary matrix, $\breve{U}$, such that $\breve{U}L\breve{U}^* = L_N^\diamond$ and $\breve{U}M\breve{U}^* = -B_N^\diamond$, compare with (2.12) and (2.13). ◆◆

11.2 Scattering coordinates

To compute the generalized free energy, we will use action–angle variables, as in Section 9.2 for the Toda fluid. This time, the map is less explicit and some stretch of linear algebra is in demand. In fact, we do start with the Lax matrix L in (11.8) for particles moving on the entire real line. Since particles cannot cross, we order them in increasing index. Then, the particle phase space is $(q, p) \in \Gamma_N^\triangleright = \mathbb{W}_N \times \mathbb{R}^N$ with the Weyl chamber $\mathbb{W}_N = \{q_1 < \cdots < q_N\}$, while the phase space of scattering coordinates is $(\lambda, \phi) \in \Gamma_N^\triangleright$ with ordered eigenvalues. The eigenvalue problem for L reads

$$L\psi_\alpha = \lambda_\alpha \psi_\alpha, \tag{11.25}$$

$\alpha = 1, \ldots, N$. Since $L = L^*$, there is a unitary matrix, U, such that

$$ULU^* = \mathrm{diag}(\lambda_1, \ldots, \lambda_N). \tag{11.26}$$

We also introduce

$$A = \mathrm{diag}(\mathrm{e}^{q_1}, \ldots, \mathrm{e}^{q_N}), \quad d_j = \mathrm{e}^{q_j/2}, \tag{11.27}$$

and correspondingly,

$$\tilde{A} = \mathrm{diag}(\mathrm{e}^{-q_1}, \ldots, \mathrm{e}^{-q_N}), \quad \tilde{d}_j = \mathrm{e}^{-q_j/2}. \tag{11.28}$$

Clearly, $A\tilde{A} = 1$. Now,

$$[A, L] = \mathrm{i}\big(|d\rangle\langle d| - A\big), \quad [\tilde{A}, L] = -\mathrm{i}\big(|\tilde{d}\rangle\langle\tilde{d}| - \tilde{A}\big). \tag{11.29}$$

Setting

$$Q = UAU^*, \quad |e\rangle = U|d\rangle, \quad \tilde{Q} = U\tilde{A}U^*, \quad |\tilde{e}\rangle = U|\tilde{d}\rangle, \tag{11.30}$$

conjugating the identities (11.29) with U, and taking the i, j matrix elements result in

$$Q_{i,j} = e_i \frac{\mathrm{i}}{\lambda_j - \lambda_i + \mathrm{i}} e_j^*, \quad \tilde{Q}_{i,j} = \tilde{e}_i \frac{\mathrm{i}}{\lambda_i - \lambda_j + \mathrm{i}} \tilde{e}_j^*. \tag{11.31}$$

Since $Q\tilde{Q} = 1$, from the diagonal matrix elements $(Q\tilde{Q})_{j,j} = 1$, one concludes

$$-e_j \tilde{e}_j^* \sum_{m=1}^{N} \frac{e_m^* \tilde{e}_m}{\left(\lambda_m - \lambda_j + \mathrm{i}\right)^2} = 1. \tag{11.32}$$

In particular, the expansion coefficients $e_j, \tilde{e}_j$ cannot vanish. From the off-diagonal matrix elements, $(Q\tilde{Q})_{i,j} = 0, i \neq j$, one infers

$$\begin{aligned}
0 &= \sum_{m=1}^{N} \frac{e_m^* \tilde{e}_m (\lambda_i - \lambda_j)}{\left(\lambda_m - \lambda_i + \mathrm{i}\right)\left(\lambda_m - \lambda_j + \mathrm{i}\right)} \\
&= \sum_{m=1}^{N} \frac{e_m^* \tilde{e}_m}{\left(\lambda_m - \lambda_i + \mathrm{i}\right)} - \sum_{m=1}^{N} \frac{e_m^* \tilde{e}_m}{\left(\lambda_m - \lambda_j + \mathrm{i}\right)}.
\end{aligned} \tag{11.33}$$

Hence, the latter sum equals some constant, c, independent of j,

$$\sum_{m=1}^{N} \frac{e_m^* \tilde{e}_m}{\left(\lambda_m - \lambda_j + \mathrm{i}\right)} = c. \tag{11.34}$$

From the Cauchy-type identity in (11.43) below, it follows that

$$e_j^* \tilde{e}_j = \mathrm{i}c V_j \tag{11.35}$$

with

$$V_j = \prod_{m=1, m\neq j}^{N} \left(1 + \frac{\mathrm{i}}{\lambda_j - \lambda_m}\right), \tag{11.36}$$

while the identity (11.44) implies

$$\langle e, \tilde{e}\rangle = \mathrm{i}cN. \tag{11.37}$$

On the other hand, $\langle e, \tilde{e} \rangle = N$ from the definition (11.30) and $c = -i$ follows. Hence,

$$e_j^* \tilde{e}_j = V_j \tag{11.38}$$

and using (11.45) one checks that Eq. (11.32) is satisfied. Together with (11.31), the final result reads

$$Q_{j,j} = |e_j|^2, \quad \tilde{Q}_{j,j} = |\tilde{e}_j|^2, \quad |e_j|^2 |\tilde{e}_j|^2 = |V_j|^2. \tag{11.39}$$

One defines a real-valued ϕ_j through

$$Q_{j,j} = |V_j| e^{\phi_j}, \tag{11.40}$$

which indeed turns out to be equal to the scattering shift ϕ_j^+ of the Calogero model, compare with (2.43). Somewhat implicitly, we have obtained the scattering map $\Phi : (\lambda, \phi) \mapsto (q, p)$. From the invariance of the trace under unitary transformations, one concludes

$$\text{tr}[A] = \sum_{j=1}^{N} e^{q_j} = \text{tr}[Q] = \sum_{j=1}^{N} |V_j| e^{\phi_j},$$

$$\text{tr}[\tilde{A}] = \sum_{j=1}^{N} e^{-q_j} = \text{tr}[\tilde{Q}] = \sum_{j=1}^{N} |V_j| e^{-\phi_j}. \tag{11.41}$$

◆◆ *Cauchy-type identities*: We consider coefficients $b_j, j = 1, \ldots, N$, satisfying the linear equations

$$\sum_{j=1}^{N} \frac{b_j}{x_j - y_i} = 1, \tag{11.42}$$

$i = 1, \ldots, N$, for some given set of coefficients $\{x_j, y_j, j = 1, \ldots, N\}$. Then, the following identities hold:

$$b_j = \frac{\prod_{m=1}^{N}(x_j - y_m)}{\prod_{m=1, m \neq j}^{N}(x_j - x_m)}, \tag{11.43}$$

$$\sum_{j=1}^{N} b_j = \sum_{j=1}^{N}(x_j - y_j), \tag{11.44}$$

$$\sum_{j=1}^{N} \frac{b_j}{(x_j - y_i)^2} = -\frac{\prod_{m=1, m \neq i}^{N}(y_i - y_m)}{\prod_{m=1}^{N}(y_i - x_m)}. \tag{11.45}$$

◆◆

11.3 Generalized free energy

While the periodized system is still integrable, there seems to be no analog of the Dumitriu–Edelman change of volume elements. On the other hand, scattering coordinates and a carefully chosen box potential can still be implemented.

To the GGE potential, we add the external potential given by

$$V_{\mathrm{ex}}(q) = \sum_{j=1}^{N} \mathrm{e}^{-\ell/2} \cosh q_j, \qquad (11.46)$$

which has a very flat bottom over the interval $[-\ell/2, \ell/2]$ and then increases exponentially with rate 1. In fact, for this particular choice, the Calogero fluid remains integrable. While this property is not used directly, it is reflected by having a simple formula for the transformed V_{ex},

$$V_{\mathrm{ex}}(q) \circ \Phi = \sum_{j=1}^{N} \mathrm{e}^{-\ell/2} Y_j \cosh \phi_j \qquad (11.47)$$

with

$$Y_j = |V_j| = \prod_{m=1, m \neq j}^{N} \left(1 + \frac{1}{(\lambda_m - \lambda_j)^2} \right)^{1/2}. \qquad (11.48)$$

Using the symmetry in q, the partition function reads

$$Z_{\mathrm{ca}, N}(\ell, V) = \int_{\Gamma_N^{\triangleright}} \mathrm{d}^N q \, \mathrm{d}^N p \exp\left(-\mathrm{tr}\big[V(L)\big] - \sum_{j=1}^{N} \mathrm{e}^{-\ell/2} \cosh q_j \right)$$

$$= \int_{\Gamma_N^{\triangleright}} \mathrm{d}^N \lambda \, \mathrm{d}^N \phi \exp\left(- \sum_{j=1}^{N} \big(V(\lambda_j) + \mathrm{e}^{-\ell/2} Y_j \cosh \phi_j \big) \right).$$

$$(11.49)$$

Since the integrand is symmetric in λ, at the expense of a factor $1/N!$, the λ-integration is extended to $\mathbb{R}^N$. The ϕ_j-integration yields the order zero modified Bessel function of the second kind K_0, see (9.35), resulting in

$$Z_{\mathrm{ca}, N}(\ell, V) = \frac{1}{N!} \int_{\mathbb{R}^N} \mathrm{d}^N \lambda \prod_{j=1}^{N} \mathrm{e}^{-V(\lambda_j)} \prod_{j=1}^{N} 2 K_0\big(2 \mathrm{e}^{-\ell/2} Y_j\big). \qquad (11.50)$$

As a consequence, the joint distribution of eigenvalues of the Lax matrix (11.8) under the normalized measure

$$Z_{ca,N}(\ell,V)^{-1} \exp\left(-\text{tr}[V(L)] - \sum_{j=1}^{N} e^{-\ell/2}\cosh q_j\right) d^N p\, d^N q \qquad (11.51)$$

is given by

$$\frac{1}{Z_{ca,N}(\ell,V)} \exp\left[-\sum_{j=1}^{N} V(\lambda_j)\right]$$

$$\times \exp\left[\sum_{j=1}^{N} \log\left(2K_0\left(2\exp\left[-\tfrac{1}{2}N\left(\nu - \frac{2}{N}\sum_{i=1,i\neq j}^{N}\log Y_j\right)\right]\right)\right)\right] \qquad (11.52)$$

relative to $(1/N!)d^N\lambda$. Note that

$$-\frac{2}{N}\log Y_j = \frac{1}{N}\sum_{i=1,i\neq j}^{N} \phi_{ca}(\lambda_j - \lambda_i) \qquad (11.53)$$

with

$$\phi_{ca}(w) = -\log\left(1 + \frac{1}{w^2}\right). \qquad (11.54)$$

As discussed in the Insert, ϕ_{ca} is indeed the scattering shift of two Calogero particles.

We arrived at an expression very similar to the Toda lattice, except for a distinct Y_j. Following the previous steps with

$$\Upsilon_j = \nu + \frac{1}{N}\sum_{i=1,i\neq j}^{N} \phi_{ca}(\lambda_j - \lambda_i), \qquad (11.55)$$

the free energy functional becomes

$$\mathcal{F}_{ca}^{\circ}(\varrho) = \nu^{-1}\int_{\mathbb{R}} dw \varrho(w)\left(V(w) - 1 + \log\varrho(w) - \log\right.$$

$$\left. \times\left(\nu + \int_{\mathbb{R}} dw' \varrho(w')\phi_{ca}(w - w')\right)\right) \qquad (11.56)$$

in case

$$\nu + \int_{\mathbb{R}} dw' \varrho(w')\phi_{ca}(w - w') > 0 \qquad (11.57)$$

and $\mathcal{F}_{ca}^{\circ}(\varrho) = \infty$ otherwise. This infinite-dimensional constraint can be ignored, just as for the Toda fluid. The variation is over all ϱ such that

$\varrho > 0$ and $\langle \varrho \rangle = 1$. Substituting $\nu^{-1}\varrho = \rho$, the free energy per unit length of the Calogero fluid is determined by the free energy functional

$$\mathcal{F}_{\text{ca}}(\rho) = \int_{\mathbb{R}} \mathrm{d}w \rho(w) \left(V(w) - 1 \right.$$

$$\left. + \log \rho(w) - \log \left(1 + \int_{\mathbb{R}} \mathrm{d}w' \rho(w') \phi_{\text{ca}}(w - w') \right) \right). \quad (11.58)$$

Supporting the generic structure of GHD, $\mathcal{F}_{\text{ca}}$ agrees with (9.42) upon substituting the scattering shift ϕ_{to} by ϕ_{ca}. Therefore, the steps from (9.43) to (9.50) can be repeated, in particular leading to the TBA equation for the Calogero fluid.

•• *Two-particle Calogero scattering shift*: For two particles, the equations of motion are

$$\ddot{q}_1(t) = \tfrac{1}{4} \cosh(\tfrac{1}{2}(q_1 - q_2)) \sinh^{-3}(\tfrac{1}{2}(q_1 - q_2)),$$

$$\ddot{q}_2(t) = -\tfrac{1}{4} \cosh(\tfrac{1}{2}(q_1 - q_2)) \sinh^{-3}(\tfrac{1}{2}(q_1 - q_2)), \quad (11.59)$$

which have to be solved with the asymptotic condition $p_1(-\infty) = p_1$, $p_2(-\infty) = p_2$, and $p_2 < p_1$. Hence, $q_1(t) < q_2(t)$. The solution can be worked out explicitly with the result

$$q_1(t) = \tfrac{1}{2}(p_1 + p_2)t - \tfrac{1}{2}q(t), \quad q_2(t) = \tfrac{1}{2}(p_1 + p_2)t + \tfrac{1}{2}q(t),$$

$$q(t) = 2\log \left[b \cosh(\gamma t) + \left(b^2 \cosh^2(\gamma t) - 1 \right)^{1/2} \right], \quad (11.60)$$

where $b^2 = 1 + (2\gamma)^{-2}$ and $\gamma = \tfrac{1}{2}(p_1 - p_2) > 0$. The large time asymptotics is given by

$$q_1(t) = \begin{cases} p_1 t - \log b, \\ p_2 t - \log b, \end{cases} \quad q_2(t) = \begin{cases} p_2 t + \log b, & t \to -\infty, \\ p_1 t + \log b, & t \to \infty. \end{cases} \quad (11.61)$$

Since quasiparticle 1 is shifted by $-2\log b$ and quasiparticle 2 by $2\log b$, we conclude that the two-particle scattering shift is given by

$$\phi_{\text{ca}}(p_1 - p_2) = -\log \left(1 + \frac{1}{(p_1 - p_2)^2} \right), \quad (11.62)$$

compare with the discussion in Section 2.3. For small momentum transfer, one recovers the Toda two-particle scattering shift $2\log|p_1 - p_2|$. Since particles do not cross each other, $\phi_{\text{ca}} < 0$, consistent with the sign for hard rods. ••

11.4 Hydrodynamic equations

This discussion can be very brief. The density current is momentum, which itself is conserved. Therefore, Eq. (9.20) is still valid. In the limit $\mu \to \infty$, the density vanishes and hence also the GGE averaged currents. Equation (9.21) still holds and as a consequence,

$$v^{\text{eff}} = \frac{\varsigma_1^{\text{dr}}}{\varsigma_0^{\text{dr}}}. \tag{11.63}$$

Of course, the dressing transformation is now defined through the scattering shift (11.62). As for the Toda fluid, in terms of the properly normalized density of states, the hydrodynamic equations read

$$\partial_t \rho_{\text{p}}(x,t;v) + \partial_x\big(v^{\text{eff}}(x,t;v)\rho_{\text{p}}(x,t;v)\big) = 0, \tag{11.64}$$

in correspondence with Eq. (9.25).

11.5 Classical Bethe equations

Bethe equations determine the eigenvalues of a quantum Hamiltonian whose eigenfunctions are obtained from employing the Bethe ansatz. In Chapter 13, we will discuss such a method using the Lieb–Liniger δ-Bose gas as a prime example. A similar structure appears for other integrable quantum many-body systems. An even wider class of integrable quantum models can be handled by the asymptotic Bethe ansatz, meaning that the Bethe ansatz holds only for large separation of particles. An example is the quantum Toda lattice covered in Chapter 14. In this section, we argue that Bethe equations can be written also for the classical Calogero fluid. Of course, this is not a microscopic foundation. Of interest is merely the structural similarity. From the Bethe equations, one deduces again the validity of TBA and thereby the DOS previously recorded.

Formally, one introduces the classical "phase shift" of the Calogero fluid as

$$\theta_{\text{ca}}(w) = -w \log(1 + w^{-2}) - 2\arctan w, \quad \theta_{\text{ca}}' = \phi_{\text{ca}}, \quad \theta_{\text{ca}}(0) = 0. \tag{11.65}$$

The Bethe equations are defined through

$$u_j = \nu N \lambda_j + \sum_{i=1}^{N} \theta_{\text{ca}}(\lambda_j - \lambda_i) \tag{11.66}$$

for $j = 1, \ldots, N$. The input is $u \in \mathbb{W}_N$ and $(\lambda_1, \ldots, \lambda_N)$ is the output. The input is weighted uniformly by the Lebesgue measure $\mathrm{d}^N u$ and the induced output is assumed to have the probability

$$\frac{1}{Z} \exp\left(-\sum_{j=1}^{N} V(\lambda_j)\right) d^N u. \tag{11.67}$$

To define a probability measure, this expression has to be restricted to the domain for which the Jacobian matrix $\{\partial u_j / \partial \lambda_i\}$, $i, j = 1, \dots, N$, is strictly positive. To find out, we differentiate (11.66) with the result

$$\frac{\partial u_j}{\partial \lambda_i} = \delta_{ij}\left(\nu N + \sum_{i=1}^{N} \phi_{ca}(\lambda_j - \lambda_i)\right) - (1 - \delta_{ij})\phi_{ca}(\lambda_j - \lambda_i)$$

$$= N\left(\Lambda_{i,j}^{\circ} + N^{-1}\Lambda_{i,j}^{\perp}\right), \tag{11.68}$$

where Λ° is the diagonal part and $\Lambda^{\perp}$ the off-diagonal one. Then, (11.67) transforms to

$$\frac{1}{Z} N^N \exp\left(-\sum_{j=1}^{N} V(\lambda_j)\right) \exp\left(\text{tr}\left[\log(\Lambda^{\circ} + N^{-1}\Lambda^{\perp})\right]\right) d^N \lambda. \tag{11.69}$$

For the exponent, we write

$$\text{tr}\left[\log\left(\Lambda^{\circ}(1 + N^{-1}(\Lambda^{\circ})^{-1}\Lambda^{\perp}))\right)\right]$$
$$= \text{tr}\left[\log \Lambda^{\circ}\right] + \text{tr}\left[\log\left(1 + N^{-1}(\Lambda^{\circ})^{-1}\Lambda^{\perp}\right)\right]. \tag{11.70}$$

Expanding the latter logarithm yields

$$\frac{1}{N}\text{tr}\left[(\Lambda^{\circ})^{-1}\Lambda^{\perp}\right] - \frac{1}{2N^2}\text{tr}\left[\left((\Lambda^{\circ})^{-1}\Lambda^{\perp}\right)^2\right]. \tag{11.71}$$

Since $(\Lambda^{\circ})^{-1}$ is diagonal, the first summand vanishes, while the second one is of order 1 since there are N^2 summands each one of order 1. The leading term is of order N relative to which the remainder is small. Within this approximation, one obtains

$$\frac{1}{Z} \exp\left(-\sum_{j=1}^{N} V(\lambda_j) + \sum_{j=1}^{N} \log\left(\nu + N^{-1}\sum_{i=1}^{N} \phi_{ca}(\lambda_j - \lambda_i)\right)\right) d^N \lambda,$$

$$\tag{11.72}$$

an expression which we have encountered already. For large N, the positivity of the Jacobian is ensured by

$$\nu + N^{-1}\sum_{i=1}^{N} \phi_{ca}(\lambda_j - \lambda_i) > 0 \tag{11.73}$$

and agreement with (11.56) is accomplished. Filling in the remaining steps requires more efforts. One hurdle is the positivity of the Jacobian at fixed N, which is connected with having a unique solution to the classical Bethe equations (11.66).

For quantum systems solvable by Bethe ansatz, upon substituting the appropriate phase shift, Eq. (11.66) stays intact. Only the input Lebesgue measure is replaced by the counting measure over integers $I_1 < \cdots < I_N$, which result from quantizing the momenta of the model with zero interaction.

11.6 Trigonometric Calogero–Moser model

For a high density of particles, the potential $V_{\rm ca}(x) = (2\sinh(x/2))^{-2}$ can be approximated by $V_{\rm cm}(x) = 1/x^2$, which leads to a fluid called rational Calogero–Moser model. Its Hamiltonian reads

$$H_{{\rm cm},N} = \sum_{j=1}^{N} \tfrac{1}{2}p_j^2 + \frac{1}{2} \sum_{i,j=1,i\neq j}^{N} \frac{1}{(q_i - q_j)^2}. \tag{11.74}$$

This model is still integrable with Lax pair

$$L_{i,j} = \delta_{ij}p_j + {\rm i}(1 - \delta_{ij})\frac{1}{q_i - q_j},$$

$$M_{i,j} = {\rm i}\delta_{ij}\left(\sum_{m=1,m\neq j}^{N} \frac{1}{(q_j - q_m)^2}\right) - {\rm i}(1 - \delta_{ij})\frac{1}{(q_i - q_j)^2}. \tag{11.75}$$

For systems in one dimension, equilibrium correlations generically have the same decay as the potential. Thus, the much emphasized locality of conservation laws can no longer be maintained. Nevertheless, as to be discussed, the Calogero–Moser model has a structure rather similar to the Toda and Calogero fluid. So as to speak, integrability dominates long-range interactions.

To start with a low key, we study the two-particle scattering. The relative motion $q(t) = q_1(t) - q_2(t)$ is governed by

$$\frac{{\rm d}^2}{{\rm d}t^2}q(t) = 4q(t)^{-3}. \tag{11.76}$$

For $v > 0$, we impose $\dot{q}(0) = 0$ and $q(0) = 2/v$, which matches with $\dot{q}(\pm\infty) = \pm v$. Then, the solution consists of two branches, one for $t > 0$ and one for $t < 0$, with

$$q(t) = \pm\sqrt{(vt)^2 + 4v^{-2}} \simeq \pm vt\left(1 + 2v^{-4}t^{-2}\right) \tag{11.77}$$

for large $|t|$. Since the first-order deviation from the linear motion decays as $|t|^{-1}$, the scattering shift of the rational model vanishes,

$$\phi_{\mathrm{cm}}(p_1 - p_2) = 0. \tag{11.78}$$

If GHD is still applicable, then the hyperbolic conservation laws governing the motion of the conserved fields would decouple and the Euler-type dynamics cannot be distinguished from the one of an ideal gas. Of course, in a local spacetime patch, positions are highly correlated and very far from a Poisson-like distribution.

To check whether such predictions are valid, one has to compute the generalized free energy. Due to long-range interactions, the notion of finite volume has some level of arbitrariness. One physical natural procedure is to start from an infinitely extended periodic configuration of spatial period ℓ, $\ell > 0$, and study the limit when $\ell \to \infty$. Since particles cannot cross, the order $q_j < q_{j+1}$ is maintained. The periodicity condition thus reads

$$q_{j+N} = q_j + \ell, \quad p_{j+N} = p_j \tag{11.79}$$

for all $j \in \mathbb{Z}$. The interaction potential turns then to the periodized rational potential

$$\sum_{m \in \mathbb{Z}} \frac{1}{(x - \ell m)^2} = \frac{\pi^2}{\ell^2 \sin^2(\pi x/\ell)} = V_{\mathrm{tcm}}(x). \tag{11.80}$$

Particles with interaction potential V_{tcm} are called trigonometric Calogero–Moser model. Each cell $[m\ell, (m+1)\ell]$ contains N particles. A single cell can be viewed as a ring of length ℓ, $\{z \in \mathbb{C} \mid |z| = \ell/2\pi\}$. If q_i, q_j are the locations of particles on the ring, then their interaction potential is $1/d(q_i, q_j)^2$, where $d(q_i, q_j)$ denotes their cord distance.

The phase space of this model can be taken as $\Gamma^{\triangleright}_{N,\ell} = \mathbb{W}_{N,\ell} \times \mathbb{R}^N$ with the Weyl chamber $\mathbb{W}_{N,\ell} = \{0 \leqslant q_1 < \cdots < q_N < \ell\}$. The trigonometric Calogero–Moser model is still integrable with Lax matrix

$$L_{i,j} = p_j \delta_{ij} + \mathrm{i}(1 - \delta_{ij}) \frac{\pi}{\ell \sin(\pi(q_i - q_j)/\ell)}, \tag{11.81}$$

which depends on ℓ, $L = L(\ell)$. Through the canonical transformation $\check{p}_j = \ell p_j$, $\check{q}_j = q_j/\ell$, one obtains the scaling with ℓ as

$$\ell L(\ell) = L(1). \tag{11.82}$$

As a side remark, for the Calogero fluid, the interaction is $V_{\mathrm{ca}}(x) = g^2(\mu/2)^2(\sinh(x/2\mu))^{-2}$. Thus, the periodized potential (11.80) corresponds to $g = 1$ and $\mu = i2\pi/\ell$, which is merely the analytic continuation of the potential V_{ca} to purely imaginary arguments.

Since particles move on the circle, rather than scattering coordinates, the conventional action–angle variables come into use. Phase space is foliated into invariant tori and the trajectories are quasi-periodic in time. The action–angle variables are constructed by an algebra very similar to the one in Section 11.2. Under the action–angle map Φ, the angle variables vary as $\phi_j \in [0, \ell]$. In comparison to the hyperbolic model, the transformation to the eigenvalues is much more explicit, in the sense that the eigenvalues are merely constrained as

$$\lambda_{j+1} - \lambda_j \geqslant 2\pi/\ell, \tag{11.83}$$

which defines the set $\Omega_{N,\ell}$. The Boltzmann weight then transforms as

$$e^{-\text{tr}[V(L(\ell))]}d^N q\, d^N p = \exp\left(-\sum_{j=1}^{N} V(\lambda_j)\right) \chi(\Omega_{N,\ell}) d^N \lambda\, d^N \phi, \tag{11.84}$$

where $\chi(\Omega_{N,\ell})$ is the characteristic function of the set $\Omega_{N,\ell}$.

We first consider the partition function, in fact, its version with the additional constraint $\lambda_N \leqslant w$. The set of admissible eigenvalues is then denoted by

$$\Omega_{N,\ell}(w) = \{\lambda \in \mathbb{R}^N \mid \lambda_{j+1} - \lambda_j \geqslant (2\pi/\ell), \quad j = 1, \ldots, N-1, \quad \lambda_N \leqslant w\} \tag{11.85}$$

with the convention $\Omega_{N,\ell} = \Omega_{N,\ell}(\infty)$. The constrained partition function is given by

$$Z_{N,\ell}(V, w) = \ell^N \int_{\Omega_{N,\ell}(w)} d^N \lambda \prod_{j=1}^{N} e^{-V(\lambda_j)}, \tag{11.86}$$

including the cases $Z_{0,\ell}(V, w) = 1$ and $Z_{1,\ell}(V, w) = \ell \int_{-\infty}^{w} d\lambda_1 \exp(-V(\lambda_1))$. The factor ℓ^N results from the ϕ-integrations. By introducing the chemical potential μ, dual to the particle number, we switch to the grand canonical ensemble as

$$Z_\ell(V, w) = \sum_{N=0}^{\infty} e^{\mu N} Z_{N,\ell}(V, w) \tag{11.87}$$

and note that

$$Z'_\ell(V, w) = \ell e^{-V(w)+\mu} Z_\ell(V, w - (2\pi/\ell)) \tag{11.88}$$

with $Z'_\ell = (\mathrm{d}/\mathrm{d}w)Z_\ell$. The constrained free energy per unit length is

$$F_\ell(V, w) = \frac{1}{\ell} \log Z_\ell(V, w), \quad \lim_{\ell \to \infty} \frac{1}{\ell} \log Z_\ell(V, w) = F(V, w). \quad (11.89)$$

Commonly, the free energy would be $-F(V, w)$. With our definition, $F'(V, w) > 0$ which will be more convenient. Inserting in (11.88), one concludes that in the infinite volume limit,

$$F'(V, w) = \mathrm{e}^{-V(w)+\mu} \mathrm{e}^{-2\pi F'(V,w)}. \quad (11.90)$$

The connection with the Toda fluid can now be established by introducing

$$F'(V) = \mathrm{e}^{-\varepsilon} = \rho_{\mathsf{n}}. \quad (11.91)$$

Then, (11.89) turns into

$$\varepsilon = V - \mu + 2\pi \mathrm{e}^{-\varepsilon}, \quad (11.92)$$

which one recognizes as the TBA equation. The T operator has the kernel

$$T(w, w') = -2\pi \delta(w - w'), \quad (11.93)$$

in accordance with zero scattering shift. Observe that for hard rods the corresponding operator T is the projection onto the constant function, while here $-T/2\pi$ is the identity matrix. The $\ell \to \infty$ limit of the free energy per unit volume is given by

$$F(V, \infty) = \int_{\mathbb{R}} \mathrm{d}w F'(V, w). \quad (11.94)$$

To complete the analogy, we have to still determine the density of states of the Lax matrix per unit volume, which amounts to the average

$$\ell^{-1} \left\langle \sum_{j=1}^{N} f(\lambda_j) \right\rangle_{V,\ell}, \quad (11.95)$$

with respect to the grand canonical density as in the partition function (11.86) at $w = \infty$. This average can be accomplished by perturbing the confining potential as $V + \delta f$. Then, to first order in δ,

$$-\int_{\mathbb{R}} \mathrm{d}w f(w) \partial_\delta F'(V, w)|_{\delta=0} = \int_{\mathbb{R}} \mathrm{d}w f(w) \rho_{\mathsf{p}}(w). \quad (11.96)$$

Using (11.90) to work out the derivative with respect to δ at $\delta = 0$ yields

$$\rho_{\mathsf{p}}(w) = \frac{F'(V, w)}{1 + 2\pi F'(V, w)}, \quad \rho_{\mathsf{s}}(w) = \frac{1}{1 + 2\pi F'(V, w)}. \quad (11.97)$$

Then, indeed $\rho_{\mathsf{p}} = \rho_{\mathsf{n}}\rho_{\mathsf{s}}$, $\rho_{\mathsf{s}} = 1 + T\rho_{\mathsf{p}}$, and $\rho_{\mathsf{n}}(w) = 1 + 2\pi F'(V, w)$. As a final step, the effective velocity can be determined through (6.20) with the result

$$v_{\text{eff}}(w) = w, \tag{11.98}$$

implying the completely decoupled set of hydrodynamic equations

$$\partial_t \rho_{\mathsf{p}}(x, t; v) + v\partial_x \rho_{\mathsf{p}}(x, t; v) = 0. \tag{11.99}$$

Notes and references

Section 11.0

The famous hierarchical model was introduced by Dyson (1969) to study ferromagnetic Ising models in one dimension with long-range interactions. Johansson (1991) established the liquid–gas transition for a one-dimensional fluid with a potential decaying as $-r^{-\alpha}$, $1 \leqslant \alpha < 2$. The functional relations for $\alpha, \beta, \gamma, V_{\text{mec}}$ are studied in Calogero (2001) and in Olshanetsky and Perelomov (1981). Interestingly enough, the construction of the Calogero soliton uses the Lax matrix $\tilde{L}$, see Gon and Kulkarni (2019).

Section 11.1

Calogero (1971) and Sutherland (1971) independently discovered the integrability of the *quantum* many-body system with $1/x^2$ interaction potential. A few years later, the classical version was investigated by Calogero (1975) as part of a wider class of classical systems having a Lax matrix pair. This class also includes the two models discussed in this chapter. Independently, Moser (1975a) discovered the Lax matrix for the rational and trigonometric model. In the rational case, he proves that the N-particle system has zero scattering shift. For the trigonometric model, he found concise expressions for the time-dependent solution. In honor of these early contributions, the names Calogero–Moser–Sutherland are attached to the models already listed and their vast generalizations discovered in the mean time, as documented in the book edited by van Diejen and Vinet (2012). We cover only two specific cases, namely the Calogero fluid, interaction potential $1/\sinh^2$, and the Calogero–Moser model, interaction potential $1/x^2$ and its periodized version $1/\sin^2$. The 2001 monograph by Calogero (2001) is a rich source on integrable classical particle models. The hyperbolic model is discussed in Calogero (2001), Section 2.1.5. The connection to Lie algebras is reviewed by Olshanetsky and Perelomov (1981). When adding a fine-tuned external potential, integrability is preserved, see the contributions by

Inozemtsev (1983), Polychronakos (1992), and Kulkarni and Polychronakos (2017). The strength can be arbitrary, but the decay has to match the one for the $1/\sinh^2$ potential. The term $-\frac{1}{12}$ appearing in (11.22) is discussed by Ruijsenaars (1999).

Section 11.2

The scattering coordinates for the hyperbolic Calogero model have been constructed by Ruijsenaars (1988). Our exposition follows the very readable contribution of Bogomolny *et al.* (2011), who provide the required algebra. The definition of the scattering shifts through (11.40) is based on analogy with earlier contributions by Airault *et al.* (1977) and Adler (1977). The difficult part is to establish agreement with the dynamical approach in (2.43). Only then the required properties of Φ can be proved, as accomplished by Ruijsenaars (1988). Ruijsenaars (1999) provides an instructive summary of his own contributions.

Section 11.3

While there is some discussion in the literature, for example, Polychronakos (2011), the variational principle for the free energy is a recent result. The two-particle scattering shift has been computed by Calogero (2001), Section 2.1.5. The Calogero fluid with external potential $g_1 \cosh x + g_2 \cosh 2x$, compare with (11.46), is still integrable, see van Diejen (1995) for a proof. But the respective Lax matrix is less accessible.

Section 11.5

Classical Bethe equations seem to be novel. To study such a problem, a convenient starting point would be to assume a toy model defined by $\phi_{ca} > 0$. Then, the Jacobian matrix is positive and Bethe equations have a unique solution.

Section 11.6

The construction of action–angle variables can be found in Ruijsenaars (1995) under model III_{nr}. Since also other models are discussed in parallel, the results of interest in the context of our discussions are scattered throughout the article. A concise account of the algebraic steps is provided by Bogomolny *et al.* (2011). Our discussion is based on the clarifying contribution by Choquard (2000), who covers also the quantized model. In the recent contribution by Bulchandani *et al.* (2021), both the quantum and classical rational Calogero–Moser models are studied, emphasizing the aspect of quasiparticles. Presented are plots of the DOS and also time-dependent solutions.

Chapter 12

Discretized Nonlinear Schrödinger Equation

12.1 Continuum wave equations

A famous integrable classical field theory in 1+1 dimensions is the nonlinear Schrödinger equation (NLS). In the defocusing case, the wave field, $\psi(x, t) \in \mathbb{C}$, is governed by

$$i\partial_t \psi = -\partial_x^2 \psi + 2|\psi|^2 \psi. \tag{12.1}$$

Our interest will be random initial conditions linked to the hydrodynamic scale. For NLS, the densities of the locally conserved fields are known. The starting entries of the list read

$$Q^{[0]}(x) = |\psi(x)|^2, \quad Q^{[1]}(x) = -i\bar{\psi}(x)\partial_x \psi(x),$$

$$Q^{[2]}(x) = |\partial_x \psi(x)|^2 + |\psi(x)|^4. \tag{12.2}$$

For a given bounded interval $\Lambda \subset \mathbb{R}$, with periodic boundary conditions, the total conserved quantities then become

$$Q_\Lambda^{[n]} = \int_\Lambda \mathrm{d}x Q^{[n]}(x), \tag{12.3}$$

$n = 0, 1, \ldots$. Formally, the time-stationary generalized Gibbs ensembles are of the form

$$\exp\left[-\sum_{n=0}^\infty \mu_n Q_\Lambda^{[n]}\right] \prod_{x \in \Lambda} \mathrm{d}^2 \psi(x), \tag{12.4}$$

where the μ_ns are suitable chemical potentials. For the thermal case, $n = 0, 1, 2$, much work has been invested to construct a proper probability measure. The basic idea of the construction is easily explained. Obviously, one can lattice discretize the volume Λ. Then, the product Lebesgue measure makes sense. However, the limit of zero lattice spacing is ill-defined within standard measure theory. On the other hand, the weighted measure

$$\frac{1}{Z_{\text{nls},N}} \prod_{j=1}^{N} dx_j \exp\left(-\mu_0 \sum_{j=1}^{N} |\psi(j)|^2 - \mu_2 \sum_{j=1}^{N} |\psi(j+1) - \psi(j)|^2\right), \quad (12.5)$$

with boundary condition $\psi(N+1) = \psi(1)$, has a well-defined continuum limit, which is Gaussian and well known as stationary $\mathbb{R}^2$-valued Ornstein–Uhlenbeck process. The remaining GGE piece reads

$$\exp\left(-\mu_2 \int_{\Lambda} dx |\psi(x)|^4\right). \quad (12.6)$$

If $\mu_2 > 0$, it is integrable with respect to the Ornstein–Uhlenbeck process. As a separate issue, it has to be shown that the so constructed measure is invariant under the NLS dynamics. Recently, such a method has been extended to a much larger class of generalized Gibbs measures in case the sum in (12.4) is restricted to some highest even n. To provide a very rough idea, the Gaussian *a priori* measure now becomes a generalized Ornstein–Uhlenbeck process with action

$$\int_{\Lambda} dx\left(|\partial_x^{n/2} \psi(x)|^2 + |\psi(x)|^2\right). \quad (12.7)$$

As for $n = 2$, the technical part is to establish that, for an appropriate choice of chemical potentials, the exponential of all remaining terms can be integrated with respect to this Gaussian measure. In essence, the thereby constructed measure is concentrated on $(n-2)$ times differentiable paths. Time-stationarity is established separately. To go beyond such an existence result seems to be a challenging problem.

Given the difficulties in even obtaining the basic building blocks, a different strategy might be pursued. In numerical simulations of NLS, one discretizes the equation. While generically this would break integrability, surprisingly enough, in many cases, there is one very specific discretization for which integrability is maintained. For NLS, such a discretization was discovered in 1975 by Ablowitz and Ladik and is usually referred to as IDNLS, i.e. integrable discrete NLS. For convenience, we will use here AL as an acronym.

Another example in the same spirit is the classical sinh-Gordon equation,

$$\partial_t^2 \phi - \partial_x^2 \phi + \sinh \phi = 0, \quad (12.8)$$

with $\phi(x, t)$ a real-valued wave field. This dynamics is integrable, while its naive discretization is not. But there is a more subtle lattice version which is still integrable. Its construction is based on an analytic continuation of the discretized sine-Gordon equation.

A further much studied classical system is the Landau–Lifshitz model of a one-dimensional magnet. Here, the spin field is a three-vector, $\vec{S}(x, t)$, with $|\vec{S}| = 1$. In the simplest case of an isotropic interaction, no external magnetic field, the continuum equations of motion are

$$\partial_t \vec{S} = \vec{S} \wedge \partial_x^2 \vec{S}, \tag{12.9}$$

which is integrable. In the naive discretization, the Hamiltonian would be

$$H_{\mathrm{ll}} = \sum_{j \in \mathbb{Z}} \vec{S}_j \cdot \vec{S}_{j+1}, \tag{12.10}$$

which is not integrable, while the integrable discrete model turns out to be governed by the Hamiltonian

$$\tilde{H}_{\mathrm{ll}} = \sum_{j \in \mathbb{Z}} \log(1 + \vec{S}_j \cdot \vec{S}_{j+1}). \tag{12.11}$$

The nonlinear Schrödinger equation and its Ablowitz–Ladik discretization are solvable through IST, compare with Section 10.1. IST is a powerful and systematic method to obtain n-soliton solutions. Generically, the one-soliton solution is still explicit, while the asymptotics of the two-soliton solution yields the scattering shift. Thus, the basic input for a soliton-based hydrodynamic scale is available. For NLS, the spectral parameter turns out to be complex, which foreshadows the possibility of spectral parameters different from $\mathbb{R}$. Here, we will pursue the in spirit particle-based scheme for the Ablowitz–Ladik lattice.

12.2 Ablowitz–Ladik discretization

Upon discretization, the wave field is over the one-dimensional lattice $\mathbb{Z}$, $\psi_j(t) \in \mathbb{C}$, and is governed by

$$i\frac{\mathrm{d}}{\mathrm{d}t}\psi_j = -\psi_{j-1} + 2\psi_j - \psi_{j+1} + |\psi_j|^2(\psi_{j-1} + \psi_{j+1}), \tag{12.12}$$

hence

$$i\frac{\mathrm{d}}{\mathrm{d}t}\psi_j = -(1 - |\psi_j|^2)(\psi_{j-1} + \psi_{j+1}) + 2\psi_j. \tag{12.13}$$

Setting $\alpha_j(t) = \mathrm{e}^{2\mathrm{i}t}\psi_j(t)$, one arrives at the standard version

$$\frac{\mathrm{d}}{\mathrm{d}t}\alpha_j = \mathrm{i}\rho_j^2(\alpha_{j-1} + \alpha_{j+1}), \quad \rho_j^2 = 1 - |\alpha_j|^2. \tag{12.14}$$

Clearly, the natural phase space is $\alpha_j \in \mathbb{D}$ with the unit disk $\mathbb{D} = \{z||z| \leqslant 1\}$. In principle, whenever $\alpha_j(t)$ hits the boundary of $\mathbb{D}$, it freezes and thereby decouples the system. As we will discuss, a conservation law ensures that, if initially away from the boundary, the solution will stay so forever.

For determining the generalized free energy, the method of scattering coordinates does not seem to be available. An alternative set-up would be the closed Ablowitz–Ladik model, which consists of a finite ring of N sites, labeled as $j = 0, \ldots, N-1$, with periodic boundary conditions, $\alpha_{j+N} = \alpha_j$. While the system with periodic boundary conditions is integrable, there seems to be no method for obtaining its generalized free energy in the limit $N \to \infty$. However, there is an analog of $T = L^\circ$ from the Toda chain, in the sense that the interactions for bonds $(-1, 0)$ and $(N-2, N-1)$ are modified. In analogy, this is called the open Ablowitz–Ladik lattice. As for the Toda chain, to compute the free energy, one has to impose a linear pressure ramp of slope $1/N$. Closed (periodic) and open chains will be discussed separately.

12.2.1 *Conserved fields*

We consider a ring of N sites. The evolution equations are of Hamiltonian form by regarding α_j and its complex conjugate $\bar{\alpha}_j$ as canonically conjugate variables and by introducing the weighted Poisson bracket

$$\{f, g\}_{\mathrm{AL}} = \mathrm{i} \sum_{j=0}^{N-1} \rho_j^2 \big(\partial_{\bar{\alpha}_j} f \partial_{\alpha_j} g - \partial_{\alpha_j} f \partial_{\bar{\alpha}_j} g\big). \tag{12.15}$$

The Hamiltonian of the Ablowitz–Ladik system then reads

$$H_{\mathrm{al},N} = -\sum_{j=0}^{N-1} \big(\alpha_{j-1}\bar{\alpha}_j + \bar{\alpha}_{j-1}\alpha_j\big). \tag{12.16}$$

One readily checks that indeed

$$\frac{\mathrm{d}}{\mathrm{d}t}\alpha_j = \{\alpha_j, H_{\mathrm{al},N}\}_{\mathrm{AL}} = \mathrm{i}\rho_j^2(\alpha_{j-1} + \alpha_{j+1}). \tag{12.17}$$

The next step is to find out the locally conserved fields. As discovered by Irena Nenciu, the Ablowitz–Ladik model has also a Lax matrix, this time in the form of a Cantero–Moral–Velázquez (CMV) matrix. The basic building blocks are 2×2 matrices, which require N to be *even* because of periodic boundary conditions. One defines

$$\Xi_j = \begin{pmatrix} \bar{\alpha}_j & \rho_j \\ \rho_j & -\alpha_j \end{pmatrix} \tag{12.18}$$

and forms the $N \times N$ matrices

$$L_N = \mathrm{diag}(\Xi_0, \Xi_2, \ldots, \Xi_{N-2}), \quad M_N = \mathrm{diag}(\Xi_1, \Xi_3, \ldots, \Xi_{N-1}), \tag{12.19}$$

where Ξ_{N-1} respects the periodic boundary conditions. More pictorially, L_N corresponds to the 2×2 blocks $(0, 1), \ldots, (N-2, N-1)$, while M_N uses 2×2 blocks shifted by one, $(1, 2), \ldots, (N-1, 0)$. The CMV matrix associated with the coefficients $\alpha_0, \ldots, \alpha_{N-1}$ is then given by

$$C_N = L_N M_N. \tag{12.20}$$

Obviously, L_N, M_N are unitary and so is C_N. The eigenvalues of C_N are denoted by $e^{i\vartheta_j}$, $\vartheta_j \in [0, 2\pi]$, $j = 1, \ldots, N$. Of course, the eigenvalues depend on N, which is suppressed in our notation.

Next, we define for a general matrix, A, the $+$ operation as

$$(A_+)_{i,j} = \begin{cases} A_{i,j} & \text{if } i < j, \\ \frac{1}{2} A_{j,j} & \text{if } i = j, \\ 0 & \text{if } i > j. \end{cases} \tag{12.21}$$

Then, one version of the Lax pair reads

$$\{C_N, \mathrm{tr}(C_N)\}_{\mathrm{AL}} = i[C_N, C_{N+}], \quad \{C_N, \mathrm{tr}(C_N^*)\}_{\mathrm{AL}} = i[C_N, (C_{N+})^*]. \tag{12.22}$$

Since the Poisson bracket acts as a derivative, one deduces

$$\{(C_N)^n, \mathrm{tr}(C_N)\}_{\mathrm{AL}} = \sum_{m=0}^{n-1} (C_N)^m i[C_N, C_{N+}](C_N)^{n-1-m} = i[(C_N)^n, C_{N+}], \tag{12.23}$$

and similarly

$$\{(C_N)^n, \mathrm{tr}(C_N^*)\}_{\mathrm{AL}} = i[(C_N)^n, (C_{N+})^*]. \tag{12.24}$$

Therefore, locally conserved fields are given by

$$Q^{[n],N} = \mathrm{tr}\big[(C_N)^n\big]. \tag{12.25}$$

Their mutual Poisson brackets vanish,

$$\{Q^{[n],N}, Q^{[m],N}\}_{\mathrm{AL}} = 0. \tag{12.26}$$

The fields can be turned real-valued by considering real and imaginary parts,

$$Q^{[n,+],N} = \tfrac{1}{2}\mathrm{tr}\big[(C_N)^n + (C_N^*)^n\big] = \mathrm{tr}\big[\cos((C_N)^n)\big],$$
$$Q^{[n,-],N} = -\tfrac{1}{2}\mathrm{i}\,\mathrm{tr}\big[(C_N)^n - (C_N^*)^n\big] = \mathrm{tr}\big[\sin((C_N)^n)\big], \tag{12.27}$$

with $n = 1, \ldots, N/2$. In particular, $H_{\mathrm{al},N} = 2Q^{[1,+]} = \mathrm{tr}[C_N + C_N^*]$. These fields have a density, respectively, given by

$$Q_j^{[n],N} = Q_j^{[n,+],N} + \mathrm{i}Q_j^{[n,-],N} = ((C_N)^n)_{j,j}. \tag{12.28}$$

Although the matrices L_N, M_N have a basic 2×2 structure, the densities of the conserved fields are shift covariant by 1 in the sense

$$Q_{j+1}^{[n,\sigma],N}(\alpha) = Q_j^{[n,\sigma],N}(\tau\alpha), \quad \sigma = \pm 1, \tag{12.29}$$

with the shift operator $(\tau\alpha)_j = \alpha_{j+1} \bmod(N)$. To confirm, one introduces the unitary shift matrix S through $(S^*AS)_{i,j} = A_{i+1,j+1} \bmod(N)$. Then,

$$S^*C_N(\alpha,\bar\alpha)S = S^*L_N(\alpha,\bar\alpha)SS^*M_N(\alpha,\bar\alpha)S$$
$$= M_N(\tau\alpha,\tau\bar\alpha)L_N(\tau\alpha,\tau\bar\alpha) = C_N(\tau\alpha,\tau\bar\alpha)^{\mathrm{T}}, \tag{12.30}$$

which implies (12.29).

Later on, we will consider the infinite volume limit, $N \to \infty$. This will always be understood as a two-sided limit. For example, the infinite volume limit of L_N, denoted by L, is $L = \mathrm{diag}(\ldots, \Xi_{-2}, \Xi_0, \Xi_2, \ldots)$ and correspondingly $M = \mathrm{diag}(\ldots, \Xi_{-1}, \Xi_1, \Xi_3, \ldots)$. L, M are unitary operators on the Hilbert space $\ell_2(\mathbb{Z})$ and so is $C = LM$. The traces in (12.25) have no limit, but densities do. The matrix elements of C^n can be expanded as the sum

$$(C^n)_{i,j} = \sum_{j_1 \in \mathbb{Z}} \cdots \sum_{j_{2n} \in \mathbb{Z}} L_{i,j_1} M_{j_1,j_2} \cdots L_{j_{2n-1},j_{2n}} M_{j_{2n},j}, \tag{12.31}$$

which consists of a finite number of terms, only. For the infinite system, the index n runs over all positive integers. The infinite volume densities, $Q_j^{[n,\sigma]}$, are strictly local functions of α with support of at most $2n$ sites. The sum in (12.31) can be viewed as resulting from a nearest neighbor $2n$ step random walk from left to right. For this purpose, one considers a checkerboard on $[0, 2n] \times \mathbb{R}$. The unit square with corners $(0,0), (1,0), (1,1), (0,1)$ is

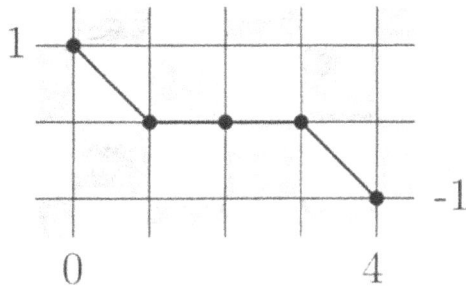

Fig. 12.1. An admissible walk from $(0, 1)$ to $(4, -1)$. According to the rules, its weight is given by $\rho_0(-\alpha_{-1})\bar{\alpha}_0\rho_{-1}$.

white. Single steps of the walk are either horizontal, $j \rightsquigarrow j$, or up-down, $j \rightsquigarrow j \pm 1$. Such diagonal steps are permitted only on white squares, compare with Figure 12.1. The matrix element $(C^n)_{i,j}$ is then the sum over all walks with $2n$ steps starting at i and ending at j. Each walk represents a particular polynomial obtained by taking the product of local weights along the walk. The weights are as follows:

ρ_j for the diagonal steps $j \rightsquigarrow j+1$ and $j+1 \rightsquigarrow j$,

$\bar{\alpha}_j$ for the horizontal step $j \rightsquigarrow j$ in case its lower square is black,

$-\alpha_{j-1}$ for the horizontal step $j \rightsquigarrow j$ in case its upper square is black.

As examples, $C_{j,j} = -\alpha_{j-1}\bar{\alpha}_j$ and $(C^2)_{j,j} = \alpha_{j-1}^2\bar{\alpha}_j^2 - \rho_{j-1}^2\alpha_{j-2}\bar{\alpha}_j - \rho_j^2\alpha_{j-1}\bar{\alpha}_{j+1}$. Note that densities are not unique in general, while the total conserved fields, $Q^{[n],N}$, are unique. To illustrate, in the previous formula, an equivalent density would be $\alpha_{j-1}^2\bar{\alpha}_j^2 - 2\rho_j^2\alpha_{j-1}\bar{\alpha}_{j+1}$.

The CMV matrix misses one physically very important field, namely

$$Q^{[0],N} = -\sum_{j=0}^{N-1} \log(\rho_j^2).\tag{12.32}$$

There is only a single index $[0]$, no distinction as $\sigma = \pm$. The time derivative of $Q^{[0],N}$ yields a telescoping sum, which vanishes on a ring. Due to lack of a common name, we call $Q^{[0],N}$ the log intensity. The log intensity vanishes for small amplitudes $|\alpha_j|^2$ and diverges at the maximal value, $|\alpha_j|^2 = 1$. Note also that

$$\exp\left(-Q^{[0],N}\right) = \prod_{j=0}^{N-1} \rho_j^2\tag{12.33}$$

is conserved. Thus, if initially $\exp\left(-Q^{[0],N}\right) > 0$, it stays so for all times, guaranteeing that the phase space boundary is never reached.

12.2.2 Generalized Gibbs ensemble

The natural *a priori* measure of the Ablowitz–Ladik model is the product measure

$$\prod_{j=0}^{N-1} d^2\alpha_j (\rho_j^2)^{P-1} = \prod_{j=0}^{N-1} d^2\alpha_j (\rho_j^2)^{-1} \exp\left(-PQ^{[0],N}\right) \qquad (12.34)$$

on $\mathbb{D}^N$. To normalize the measure, $P > 0$ is required. The log intensity is controlled by the parameter P which, in analogy to the Toda lattice, is called pressure. Small P corresponds to maximal log intensity, i.e., $|\alpha_j|^2 \to 1$, and large P to low log intensity, i.e., $|\alpha_j|^2 \to 0$. In the grand-canonical ensemble, the Boltzmann weight is constructed from a linear combination of the conserved fields, which is written as

$$\sum_{n\in\mathbb{Z}} \mu_n \mathrm{tr}\left[(C_N)^n\right] = \mathrm{tr}\left[V(C_N)\right], \quad V(z) = \sum_{n\in\mathbb{Z}} \mu_n z^n. \qquad (12.35)$$

The chemical potentials, μ_n, are complex and assumed to be independent of N. To have the trace real-valued, one imposes $\mu_n = \bar{\mu}_{-n}$. In fact is suffices to assume that V is real-valued and continuous on the circle $|z| = 1$. Combining (12.34) and (12.35) yields the generalized Gibbs ensemble as

$$\frac{1}{Z_{\mathrm{al},N}(P,V)} \prod_{j=0}^{N-1} d^2\alpha_j (\rho_j^2)^{P-1} \exp\left(-\mathrm{tr}[V(C_N)]\right). \qquad (12.36)$$

$Z_{\mathrm{al},N}(P,V)$ is the normalizing partition function. As label of the GGE, the more natural object turns out to be the Fourier transform of the sequence $\{\mu_n, n \in \mathbb{Z}\}$,

$$V(w) = V(e^{iw}). \qquad (12.37)$$

V is a real-valued continuous function on $[0, 2\pi]$. The corresponding object for the Toda lattice lives on $\mathbb{R}$ and is called confining potential, since it confines the eigenvalues of the Lax matrix. While for the CMV matrix, the eigenvalues are on the unit circle, hence there is nothing to confine; for convenience, we still stick to "confining" so to distinguish from other potentials. If V is given by a finite sum, then the interaction of the Gibbs measure in (12.36) is of finite range. In this case, the infinite volume limit can be controlled through transfer matrix methods. In particular, the limit measure is expected to have a finite correlation length. As before, finite volume expectations with respect to the measure in (12.36) are denoted by $\langle \cdot \rangle_{P,V,N}$ and their infinite volume limit by $\langle \cdot \rangle_{P,V}$.

The generalized free energy, F_{al}, is defined through

$$\lim_{N \to \infty} -\frac{1}{N} \log Z_{\text{al},N}(P,V) = F_{\text{al}}(P,V). \qquad (12.38)$$

In the hydrodynamic context of particular interest is the empirical density of states,

$$\rho_{Q,N}(w) = \frac{1}{N} \sum_{j=1}^{N} \delta(w - \vartheta_j), \qquad (12.39)$$

where $\{e^{i\vartheta_j}\}$ are the eigenvalues of C_N. $\rho_{Q,N}$ is a probability measure on $[0, 2\pi]$ and has an almost sure limit as

$$\lim_{N \to \infty} \rho_{Q,N}(w) = \rho_Q(w). \qquad (12.40)$$

To grasp the significance of the DOS, we first introduce the trigonometric functions $\varsigma_0(w) = 1$, $\varsigma_{n-}(w) = \sin(nw)$, and $\varsigma_{n+}(w) = \cos(nw)$, $n = 1, 2, \ldots$. They span the Hilbert space $L^2([0, 2\pi], dw)$. Then, for the trigonometric moments of $\rho_{Q,N}(w)$,

$$\langle \rho_{Q,N} \varsigma_{n\sigma} \rangle = N^{-1} \langle Q^{[n,\sigma],N} \rangle_{P,V,N} \qquad (12.41)$$

and

$$\lim_{N \to \infty} N^{-1} \langle Q^{[n,\sigma],N} \rangle_{P,V,N} = \langle Q_0^{[n,\sigma]} \rangle_{P,V} = \langle \rho_Q \varsigma_{n\sigma} \rangle. \qquad (12.42)$$

Consistent with our previous notation, $\langle \cdot \rangle$ is simply a short hand for the integration over $[0, 2\pi]$. The limit value can also be expressed as variational derivative of the generalized free energy per site,

$$\frac{d}{d\kappa} F_{\text{al}}(P, V + \kappa \varsigma_{n\sigma}) \Big|_{\kappa=0} = \langle Q_0^{[n,\sigma]} \rangle_{P,V}. \qquad (12.43)$$

In addition, one introduces the average log intensity, denoted by ν, for which

$$\nu = \langle Q_0^{[0]} \rangle_{P,V} = \partial_P F_{\text{al}}(P,V). \qquad (12.44)$$

Note that by definition $\nu > 0$. For $V = 0$, one readily confirms $F_{\text{al}}(P, 0) = \log(P/\pi)$ with log intensity $\nu(P) = P^{-1} > 0$. Hence, there is no high pressure phase as encountered for the Toda lattice, compare with the Insert in Section 9.1.

12.3 Circular random matrices with pressure ramp

We modify the CMV matrix at its two boundaries. As before, the number N of sites is even. L_N remains unchanged and M_N is modified to $M_N^\diamond$, where $(M_N^\diamond)_{0,0} = 1$, $(M_N^\diamond)_{0,N-1} = 0$, $(M_N^\diamond)_{N-1,0} = 0$, and $(M_N^\diamond)_{N-1,N-1} = \mathrm{e}^{\mathrm{i}\phi}$, $\phi \in [0, 2\pi]$. This leads to the particular CMV matrix

$$C_N^\diamond = L_N M_N^\diamond. \tag{12.45}$$

For the *a priori* measure (12.34), the pressure P is constant. To arrive at a tractable expression, the constant pressure is replaced by a linearly changing pressure with a yet arbitrary slope $-\frac{1}{2}\beta$, $\beta > 0$, according to

$$\mathrm{d}\phi \prod_{j=0}^{N-2} (\rho_j^2)^{-1}(\rho_j^2)^{\beta(N-1-j)/2}\mathrm{d}^2\alpha_j. \tag{12.46}$$

Surprisingly, Killip and Nenciu succeeded to compute the joint distribution of eigenvalues of $C_N^\diamond$ under this measure. We define the Vandermonde determinant

$$\Delta(z_1, \ldots, z_N) = \prod_{1 \leqslant i < j \leqslant N} (z_j - z_i). \tag{12.47}$$

Denoting the eigenvalues of $C_N^\diamond$ by $\mathrm{e}^{\mathrm{i}\vartheta_1}, \ldots, \mathrm{e}^{\mathrm{i}\vartheta_N}$, their joint (not normalized) distribution under the measure in (12.46) is given by

$$\tilde{\zeta}_N(\beta)|\Delta(\mathrm{e}^{\mathrm{i}\vartheta_1}, \ldots, \mathrm{e}^{\mathrm{i}\vartheta_N})|^\beta \prod_{j=1}^{N} \mathrm{d}\vartheta_j, \tag{12.48}$$

where the prefactor turns out to equal

$$\tilde{\zeta}_N(\beta) = 2^{(1-N)}\frac{1}{N!}\frac{\Gamma(\beta/2)^N}{\Gamma(N\beta/2)}. \tag{12.49}$$

Since β is a free parameter, one can choose specifically

$$\beta = \frac{2P}{N}. \tag{12.50}$$

Now, the ramp has slope $-P/N$ and, in the limit $N \to \infty$, close to the lattice point $(1 - u)N$, $0 < u < 1$, the measure of (12.46) will converge to the product measure of (12.34) with pressure uP. Since

$$\mathrm{tr}\big[V(C_N^\diamond)\big] = \sum_{j=1}^{N} V(\vartheta_j), \tag{12.51}$$

the Boltzmann weight can be naturally included in (12.46). Hence, the partition function of the system with boundary conditions turns into

$$Z_{kn,N}(P,V) = \int_{[0,2\pi]^{N-1}} \prod_{j=0}^{N-2} d^2\alpha_j \int_0^{2\pi} d\phi \prod_{j=0}^{N-2} (\rho_j^2)^{-1}(\rho_j^2)^{P(N-1-j)/N}$$

$$\times \exp\left(-\operatorname{tr}[V(C_N^\diamond)]\right) = \zeta_N(P) \int_0^{2\pi} d\vartheta_1 \cdots \int_0^{2\pi} d\vartheta_N$$

$$\times \exp\left(-\sum_{j=1}^N V(\vartheta_j) + P\frac{1}{N} \sum_{i,j=1,i\neq j}^N \log|e^{i\vartheta_i} - e^{i\vartheta_j}|\right)$$

(12.52)

with $\zeta_N(P) = \tilde\zeta_N(2P/N)$. In statistical mechanics, the probability distribution

$$\frac{1}{Z_{\log,N}} \exp\left(-\sum_{j=1}^N V(\vartheta_j) + P\frac{1}{N} \sum_{i,j=1,i\neq j}^N \log|e^{i\vartheta_i} - e^{i\vartheta_j}|\right)$$

(12.53)

is known as *circular log gas*. Since the coupling strength is proportional to $1/N$, it is the mean-field version of the log gas.

The Ablowitz–Ladik model with linearly varying pressure has a free energy per site defined through

$$F_{kn}(P,V) = \lim_{N\to\infty} -\frac{1}{N} \log Z_{kn,N}(P,V).$$

(12.54)

Since the pressure ramp has slope $1/N$, in the limit, free energies merely add up as

$$F_{kn}(P,V) = \int_0^1 du\, F_{al}(uP,V).$$

(12.55)

Before studying the infinite volume free energy, we remark that the CMV matrix $C_N^\diamond$ is still linked to a suitably modified Ablowitz–Ladik dynamics governed by the Hamiltonian

$$H_N^\diamond = \operatorname{tr}\left[C_N^\diamond + C_N^{\diamond*}\right].$$

(12.56)

Working out the Poisson brackets leads to the evolution equation

$$\frac{d}{dt}\alpha_j = i\rho_j^2(\alpha_{j-1} + \alpha_{j+1}),$$

(12.57)

$j = 0,\dots,N-2$, with the boundary conditions $\alpha_{-1} = -1$ and $\alpha_{N-1} = e^{i\phi}$. As before, $\operatorname{tr}\left[(C_N^\diamond)^n\right]$ is preserved under the dynamics. However, the *a priori* measure (12.46) is no longer stationary. The long-time dynamics on a ring differs qualitatively from the one with open boundary conditions.

The prefactor in (12.52) can be easily handled with the result

$$\lim_{N\to\infty} -\frac{1}{N} \log \zeta_N(P) = \log(2P) - 1. \tag{12.58}$$

The sum in the exponential of (12.53) has a term involving the confining potential V, which is linear in the empirical measure, ϱ_N, of the $\{\vartheta_j\}$s. The double sum of the interaction is quadratic in ϱ_N. Thus, the limiting free energy is determined by a variational principle. We first define the free energy functional

$$\mathcal{F}_{\mathrm{kn}}^{\circ}(\varrho) = \int_0^{2\pi} \mathrm{d}w \varrho(w) \left(V(w) + \log \varrho(w) \right.$$

$$\left. + \log P - P \int_0^{2\pi} \mathrm{d}w' \log |e^{\mathrm{i}w} - e^{\mathrm{i}w'}| \varrho(w') \right). \tag{12.59}$$

This functional has to be minimized over all densities ϱ, with the constraints $\varrho(w) \geqslant 0$ and $\langle \varrho \rangle = 1$. Denoting by $\varrho^\star$ the unique minimizer, one arrives at

$$F_{\mathrm{kn}}(P, V) = \log 2 - 1 + \mathcal{F}_{\mathrm{kn}}^{\circ}(\varrho^\star) \tag{12.60}$$

and thus, using (12.55),

$$F_{\mathrm{al}}(P, V) = \partial_P(P F_{\mathrm{kn}}(P, V)). \tag{12.61}$$

As for the Toda lattice, it turns out to be more convenient to absorb P into ϱ by setting $\rho = P\varrho$. Then, $P\mathcal{F}_{\mathrm{kn}}^{\circ}(P^{-1}\rho) = \mathcal{F}_{\mathrm{kn}}(\rho) - P \log P$ with the transformed free energy functional

$$\mathcal{F}_{\mathrm{kn}}(\rho) = \int_0^{2\pi} \mathrm{d}w \rho(w) \left(V(w) + \log \rho(w) - \int_0^{2\pi} \mathrm{d}w' \log |e^{\mathrm{i}w} - e^{\mathrm{i}w'}| \rho(w') \right). \tag{12.62}$$

$\mathcal{F}_{\mathrm{kn}}$ has to be minimized under the constraint

$$\rho(w) \geqslant 0, \quad \int_0^{2\pi} \mathrm{d}w \rho(w) = P \tag{12.63}$$

with minimizer denoted by $\rho^\star$. Then,

$$F_{\mathrm{al}}(P, V) = \partial_P \mathcal{F}_{\mathrm{kn}}(\rho^\star) - 1 + \log 2. \tag{12.64}$$

The constraint (12.63) is removed by introducing the Lagrange multiplier μ as

$$\mathcal{F}_{\mathrm{kn}}^{\bullet}(\rho) = \mathcal{F}_{\mathrm{kn}}(\rho) - \mu \int_0^{2\pi} \mathrm{d}w \rho(w). \tag{12.65}$$

The unique minimizer of $\mathcal{F}_{\mathrm{kn}}^{\bullet}(\rho)$ is denoted by $\rho_{\mathrm{n},\mu}$ and determined as a solution of the TBA equation

$$V(w) - \mu - 2\int_0^{2\pi} dw' \log|e^{iw} - e^{iw'}|\rho_{\mathrm{n},\mu}(w') + \log \rho_{\mathrm{n},\mu}(w) = 0. \quad (12.66)$$

The parameter μ has to be adjusted such that

$$P = \int_0^{2\pi} dw\rho_{\mathrm{n},\mu}(w). \quad (12.67)$$

To obtain the Ablowitz–Ladik free energy, we differentiate

$$\partial_P \mathcal{F}_{\mathrm{kn}}(\rho^{\star}) = \int_0^{2\pi} dw\partial_P \rho^{\star}(w)\left(V(w) + \log \rho^{\star}(w)\right.$$

$$\left. - 2\int_0^{2\pi} dw'\rho^{\star}(w')\log|e^{iw} - e^{iw'}|\right) + 1. \quad (12.68)$$

Integrating (12.66) against $\partial_P\rho^{\star}$, one arrives at

$$\partial_P \mathcal{F}_{\mathrm{kn}}(\rho_{\mathrm{n},\mu}) = \mu + 1 \quad (12.69)$$

and thus

$$F_{\mathrm{al}}(P, V) = \mu(P, V) + \log 2. \quad (12.70)$$

The Ablowitz–Ladik lattice has the property that its free energy is determined through an explicit variational problem. At this stage, a structure comparable to the Toda lattice has been reached. The modifications are minimal. The integration is over the interval $[0, 2\pi]$ corresponding to the eigenvalues e^{iw}. The basis functions are $\varsigma_0(w) = 1$, $\varsigma_{n-}(w) = \sin(nw)$, and $\varsigma_{n+}(w) = \cos(nw)$ and the kernel of the operator T is adjusted to

$$Tf(w) = 2\int_0^{2\pi} dw' \log|2\sin((w - w')/2)|f(w') \quad (12.71)$$

with $w \in [0, 2\pi]$. Under such modifications, the results in Sections 3.3 and 3.4 hold *ad verbatim*. In particular, under the GGE in (12.36), one obtains the $N \to \infty$ limit of the DOS as

$$\rho_Q = \partial_P \rho_{\mathrm{n}}. \quad (12.72)$$

We arrived at a somewhat perplexing conclusion. Clearly, the structure discussed very strongly resembles the one obtained for the Toda lattice. Our scheme is particle-based. Nevertheless, so far we did not identify an underlying particle structure. Their potential two-particle scattering shift arrives only indirectly through the generalized free energy. For the currents, we will use the availability of a conserved current and thus bypass the collision rate ansatz.

12.4 Average currents

As a fairly general fact, the average currents are related to fluctuations in the density of states. The same strategy will be pursued for the Ablowitz–Ladik lattice.

For the infinite lattice, the conserved fields satisfy a continuity equation of the form

$$\frac{\mathrm{d}}{\mathrm{d}t} Q_j^{[n,\sigma],N} = \{Q_j^{[n,\sigma],N}, H_{\mathrm{al},N}\}_{\mathrm{AL}} = J_j^{[n,\sigma],N} - J_{j+1}^{[n,\sigma],N}. \tag{12.73}$$

It will be convenient to work with the complex currents

$$J_j^{[n],N} = J_j^{[n,+],N} + \mathrm{i}J_j^{[n,-],N}. \tag{12.74}$$

Using Eq. (12.22) together with the fact that the matrix C_+ has nonvanishing matrix elements only for $(j, j+\ell)$ with $\ell = 0, 1, 2$, the time derivative is obtained as

$$\{(C^n)_{j,j}, H_{\mathrm{al}}\}_{\mathrm{AL}} = \sum_{\ell=1,2} \mathrm{i}\Big(C_{j-\ell,j}(C^n)_{j,j-\ell} - C_{j,j+\ell}(C^n)_{j+\ell,j}$$

$$+ \bar{C}_{j,j+\ell}(C^n)_{j,j+\ell} - \bar{C}_{j-\ell,j}(C^n)_{j-\ell,j}\Big). \tag{12.75}$$

The term with $\ell = 0$ does not contribute. The remaining terms look like a shift difference, but it is not, since the off-diagonal matrix elements are only two-periodic.

For $n = 1$, one finds

$$\{C_{j,j}, \mathrm{tr}(C)\}_{\mathrm{AL}} = \mathrm{i}\big(-\rho_{j-1}^2 \alpha_{j-2}\bar{\alpha}_j + \rho_j^2 \alpha_{j-1}\bar{\alpha}_{j+1}\big), \tag{12.76}$$

while

$$\{C_{j,j}, \mathrm{tr}(C^*)\}_{\mathrm{AL}} = \mathrm{i}\big(\rho_j^2 \alpha_{j+1}\bar{\alpha}_{j+1} - \rho_{j-1}^2 \alpha_{j-2}\bar{\alpha}_{j-2} + \rho_j^2\rho_{j+1}^2 - \rho_{j-2}^2\rho_{j-1}^2\big)$$

$$= \mathrm{i}\big(\alpha_{j-1}\bar{\alpha}_{j-1} - \alpha_j\bar{\alpha}_j\big) \tag{12.77}$$

for even j and

$$\{C_{j,j}, \mathrm{tr}(C^*)\}_{\mathrm{AL}} = \mathrm{i}\big(\rho_j^2 \alpha_{j-1}\bar{\alpha}_{j-1} - \rho_{j-1}^2 \alpha_j\bar{\alpha}_j\big) = \mathrm{i}\big(\alpha_{j-1}\bar{\alpha}_{j-1} - \alpha_j\bar{\alpha}_j\big) \tag{12.78}$$

for odd j. Adding the two terms, the first current reads

$$J_j^{[1]} = \mathrm{i}\big(-\rho_{j-1}^2 \alpha_{j-2}\bar{\alpha}_j + \alpha_{j-1}\bar{\alpha}_{j-1}\big). \tag{12.79}$$

For higher n, it seems to be difficult to guess the cancelations leading to an explicit difference form. But for the moment, we only need the property of currents being local, which can be ensured by an abstract argument.

♦♦ *Existence of currents*: We want to establish that there is a local current, $J_j^{[n]}$, such that

$$\{(C^n)_{j,j}, H_{\mathrm{al}}\}_{\mathrm{AL}} = J_j^{[n]} - J_{j+1}^{[n]}. \tag{12.80}$$

For this purpose, we consider a ring of size N and fix n such that $N > 4n$. For clarity, the index N is suppressed in our notation. $\{Q_0^{[n]}, H_{\mathrm{al}}\}_{\mathrm{AL}}$ is a polynomial of degree $2n + 2$. This polynomial is decomposed into patterns. A pattern consists of a specific monomial together with a collection of some of its translates. By definition, patterns are distinct. The monomial defining a pattern is denoted by ω_1 and its translates by ω_j. Then, understood modulo N, $\{Q_0^{[n]}, H_{\mathrm{al}}\}_{\mathrm{AL}}$ is a sum of terms of the form

$$\sum_{|\ell| \leqslant 2n+2} a^{(\ell)} \omega_\ell \tag{12.81}$$

with some complex coefficients $a^{(\ell)}$, which may vanish. Since $\{Q^{[n]}, H_{\mathrm{al}}\}_{\mathrm{AL}} = 0$, shift invariance implies that for every pattern

$$0 = \sum_{j=0}^{N-1} \sum_{|\ell| \leqslant 2n+2} a^{(\ell)} \omega_{\ell+j} = \sum_{|\ell| \leqslant 2n+2} a^{(\ell)} \left(\sum_{j=0}^{N-1} \omega_{\ell+j} \right), \tag{12.82}$$

and as a consequence

$$\sum_{|\ell| \leqslant 2n+2} a^{(\ell)} = 0. \tag{12.83}$$

Eliminating the non-zero $a^{(\ell)}$ with largest index, one obtains unique coefficients $b^{(\ell)}$ such that

$$\sum_{|\ell| \leqslant 2n+2} b^{(\ell)} (\omega_\ell - \omega_{\ell+1}) = J_0^\omega - J_1^\omega. \tag{12.84}$$

For the actual local current, one still has to sum over all patterns. ♦♦

To compute the GGE averaged currents, one first notes that

$$J_j^{[0]} = 2Q_j^{[1,-]}. \tag{12.85}$$

Hence, the path outlined in Section 6.1 can be followed. Omitting all intermediate steps, the result is

$$\partial_P \big(\langle J_0^{[n,\sigma]} \rangle_{P,V} + 2P \langle \varsigma_{1-}, C^\sharp \varsigma_{n\sigma} \rangle \big) = 0, \tag{12.86}$$

implying that the round bracket has to be independent of P, in particular equal to its value at $P = 0$. Since $\langle \varsigma_{1-}, C^\sharp \varsigma_{n\sigma} \rangle$ is bounded in P, the second summand vanishes at $P = 0$. For the first summand, one notes that in the limit $P \to 0$, for each j, the *a priori* measure (12.34) becomes uniform on the unit circle. Hence, $\langle \rho_j^2 \rangle_P \to 0$ as $P \to 0$ and the CMV matrix turns diagonal. Denoting $\alpha_j = e^{i\phi_j}$, $\phi_j \in [0, 2\pi]$, in the limit $P \to 0$, the GGE (12.36) converges to

$$\frac{1}{Z_{\mathrm{al},N}(0,V)} \prod_{j=0}^{N-1} \mathrm{d}\phi_j \exp\left(-\sum_{j=0}^{N-1} V(\phi_{j+1} - \phi_j) \right) \tag{12.87}$$

with boundary condition $\phi_N = \phi_0$. According to (12.79), one observes that $\langle J_0^{[1]} \rangle_{0,V} = \mathrm{i}$. By a somewhat lengthy computation, $\langle J_0^{[2]} \rangle_{0,V} = 0$. To extend the average to general n seems to be difficult, since a sufficiently explicit formula for $J_0^{[n]}$ is missing. Let us then assume that $\langle J_0^{[n]} \rangle_{0,V} = d_n$ with some constant d_n independent of V. Next, we substitute $P\varrho^* = \rho_{\mathsf{n}}$ with the result

$$P\langle \varsigma_{1-}, C^\sharp \varsigma_{n\sigma} \rangle = \langle \varsigma_{1-}, (1 - \rho_{\mathsf{n}} T)^{-1} \rho_{\mathsf{n}} \varsigma_{n\sigma} \rangle$$
$$-\nu \langle \varsigma_{1-}, (1 - \rho_{\mathsf{n}} T)^{-1} \rho_{\mathsf{n}} \rangle \langle \varsigma_{n\sigma}, (1 - \rho_{\mathsf{n}} T)^{-1} \rho_{\mathsf{n}} \rangle. \tag{12.88}$$

Noting that $(1 - \rho_{\mathsf{n}} T)^{-1} \rho_{\mathsf{n}}$ is a symmetric operator, one finally arrives at

$$\langle J_0^{[n,\sigma]} \rangle_{P,V} - d_n = -2\big(\langle \rho_{\mathsf{n}} \varsigma_{1-}^{\mathrm{dr}} \varsigma_{n\sigma} \rangle - q_{1-} \langle \rho_{\mathsf{p}} \varsigma_{n\sigma} \rangle \big) \tag{12.89}$$

with

$$q_{1-} = \langle Q_0^{[1,-]} \rangle_{P,V}. \tag{12.90}$$

The spectral function for the energy is $E(w) = 2\cos w$, hence $E'(w) = -2\sin w$. To emphasize the correspondence with the Toda lattice, see Eqs. (6.20) and (6.17), we define the effective velocity through

$$v^{\mathrm{eff}} = \frac{(-2\varsigma_{1-})^{\mathrm{dr}}}{\varsigma_0^{\mathrm{dr}}}, \quad \tilde{q}_{1-} = -2q_{1-}. \tag{12.91}$$

Then,

$$\langle J_0^{[0]} \rangle_{P,V} = -\tilde{q}_{1-}, \quad \langle J_0^{[n,\sigma]} \rangle_{P,V} - d_n = \langle (v^{\mathrm{eff}} - \tilde{q}_{1-}) \rho_{\mathsf{p}} \varsigma_{n\sigma} \rangle. \tag{12.92}$$

12.5 Hydrodynamic equations

On the hydrodynamic scale, the local GGE is characterized by the log intensity ν and the CMV density of states $\nu \rho_{\mathsf{p}}$, both of which now become spacetime-dependent. Merely inserting the average currents, and since d_n

is assumed to be a constant, one arrives at the Euler-type hydrodynamic evolution equations,

$$\partial_t \nu(x,t) - \partial_x \tilde{q}_{1-}(x,t) = 0,$$

$$\partial_t\big(\nu(x,t)\rho_\mathsf{p}(x,t;v)\big) + \partial_x\big((v^{\mathrm{eff}}(x,t;v) - \tilde{q}_{1-}(x,t))\rho_\mathsf{p}(x,t;v)\big) = 0. \quad (12.93)$$

Equation (12.93) is based on the assumption of local GGE. To actually establish such an equation from the underlying Ablowitz–Ladik lattice seems to be a difficult task.

As a general feature of generalized hydrodynamics, the equations can be transformed explicitly to a quasilinear form through

$$\rho_\mathsf{n} = \rho_\mathsf{p}(1 + (T\rho_\mathsf{p}))^{-1}, \quad (12.94)$$

compare with (6.24). Then, Eq. (12.93) turns into the normal form

$$\partial_t\rho_\mathsf{n} + \nu^{-1}(v^{\mathrm{eff}} - \tilde{q}_{1-})\partial_x\rho_\mathsf{n} = 0. \quad (12.95)$$

Since $\nu > 0$, no apparent singularities are encountered.

For the defocusing discrete NLS in one dimension, we established the form of the hydrodynamic equations. As a novel feature, their structure is determined by the mean-field version of the log gas corresponding to CUE random matrices. Thus, formally expanding the circle to a line, one recovers the hydrodynamic equations of the Toda lattice. Our analysis is pretty much on the same level as the one for the Toda lattice. Only the handling of average currents in the limit $P \to 0$ is incomplete. We hope to return to this point in the future.

12.6 Modified Korteweg–de Vries equation

Instead of the real part of $Q^{[1],N}$, one can also consider its imaginary part as Hamiltonian,

$$H_{\mathrm{kv},N} = -\mathrm{i}\sum_{j=0}^{N-1}\big(\alpha_{j-1}\bar{\alpha}_j - \bar{\alpha}_{j-1}\alpha_j\big) = -\mathrm{i}\,\mathrm{tr}\big[C_N - C_N^*\big]. \quad (12.96)$$

Then,

$$\frac{\mathrm{d}}{\mathrm{d}t}\alpha_j = \{\alpha_j, H_{\mathrm{kv},N}\}_{\mathrm{AL}} = \rho_j^2(\alpha_{j+1} - \alpha_{j-1}), \quad (12.97)$$

which is known as the Schur flow. Through a formal Taylor expansion, one argues that the continuum limit of Eq. (12.97) yields the modified

Korteweg–de Vries equation,

$$\partial_t u = \partial_x^3 u - 6u^2 \partial_x u, \tag{12.98}$$

which is a good reason to briefly touch upon (12.97).

As before, $\alpha_j \in \mathbb{D}$ and the conservation laws remain unchanged. However, the currents have to be modified from $J^{[n]}$ to $\breve{J}^{[n]}$. For the log intensity current, one obtains

$$\breve{J}_j^{[0],N} = 2Q_j^{[1,+],N} = H_{\text{al},N,j}. \tag{12.99}$$

Thus, the roles of $1-$ and $1+$ are interchanged. The arguments in Section 12.4 can be repeated without modifications. In the hydrodynamic equations (12.92), q_{1-} is replaced by q_{1+} and the effective velocity becomes

$$v^{\text{eff}} = \frac{(-2\varsigma_{1+})^{\text{dr}}}{\varsigma_0^{\text{dr}}}. \tag{12.100}$$

These considerations badly miss that the wave field of the modified Korteweg–de Vries equation can be taken as real-valued. If the initial α_j are real, then according to the evolution equation (12.97) this property is preserved in time. From the perspective of GGE, such initial conditions amount to a set of measure zero and one has to reconsider the analysis. Fortunately, the relevant transformation formula has been proved already by Killip and Nenciu. To avoid duplication of symbols, in the remainder of this section, $\bar{\alpha}_j = \alpha_j$ everywhere and hence $\alpha_j \in [-1, 1]$. The equations of motion read

$$\frac{\mathrm{d}}{\mathrm{d}t}\alpha_j = \rho_j^2(\alpha_{j+1} - \alpha_{j-1}), \tag{12.101}$$

where $\rho_j^2 = 1 - \alpha_j^2$ as before. While an obvious Hamiltonian structure is lost, one readily checks that the *a priori* measure

$$\prod_{j=0}^{N-1} \mathrm{d}\alpha_j (\rho_j^2)^{P-1} = \prod_{j=0}^{N-1} \mathrm{d}\alpha_j (\rho_j^2)^{-1} \exp\left(-PQ^{[0],N}\right) \tag{12.102}$$

is still stationary under the dynamics. The densities of the conserved fields are

$$Q_j^{[m]} = (C^m)_{j,j}, \tag{12.103}$$

$m = 1, 2, \ldots$, as before. But the imaginary part is dropped and there is no longer a distinction of $\pm$. In particular,

$$Q_j^{[0]} = -\log \rho_j^2, \quad Q_j^{[1]} = -\alpha_{j-1}\alpha_j,$$

$$Q_j^{[2]} = \alpha_{j-1}^2\alpha_j^2 - \rho_{j-1}^2\alpha_{j-2}\alpha_j - \rho_j^2\alpha_{j-1}\alpha_{j+1}. \tag{12.104}$$

For the log intensity current, $\breve{J}_j^{[0]} = 2Q_j^{[1]}$. Thus, $\breve{J}^{[0]}$ is conserved and the Toda strategy is still applicable.

Since C_N is now a real matrix, its eigenvalues come in pairs. If $e^{i\vartheta}$ is an eigenvalue, so is $e^{-i\vartheta}$. For a system of size N, there are only $\bar{N} = N/2$ independent eigenvalues, denoted by $e^{i\vartheta_1}, \ldots, e^{i\vartheta_{\bar{N}}}$. Rather than using a DOS reflecting such symmetry, it is more convenient to restrict the phase as $0 \leqslant \vartheta_j \leqslant \pi$. The empirical DOS is given by

$$\rho_{Q,\bar{N}}(w) = \frac{1}{\bar{N}} \sum_{j=1}^{\bar{N}} \delta(w - \vartheta_j), \tag{12.105}$$

$w \in [0, \pi]$, where we switched to $\bar{N}$ as the size parameter. In the limit $\bar{N} \to \infty$, $\rho_{Q,\bar{N}}(w)$ converges to the deterministic limit $\rho_Q(w)$. The GGE expectations are then

$$\langle Q_0^{[n]} \rangle_{P,V} = 2 \int_0^\pi \mathrm{d}w \rho_Q(w)\varsigma_n(w), \tag{12.106}$$

where $\varsigma_n(w) = \cos(nw)$, and the confining potential becomes

$$V(w) = \sum_{n=1}^\infty \mu_n \cos(nw) \tag{12.107}$$

with real chemical potentials μ_n. The confining potential is assumed to be continuous.

Under the measure

$$\prod_{j=0}^{N-2} (1-\alpha_j^2)^{-1}(1-\alpha_j^2)^{\beta(N-j-1)/4}(1-\alpha_j)^{a+1-(\beta/4)}(1+(-1)^j\alpha_j)^{b+1-(\beta/4)}\mathrm{d}\alpha_j \tag{12.108}$$

on $[-1, 1]^{N-1}$ with $\beta > 0$ and $a, b > -1 + (\beta/4)$, the joint (not normalized) distribution of the eigenvalues of C_N, imposing $\alpha_{N-1} = 1$, is given by

$$\tilde{\zeta}_{\bar{N}}(\beta)2^\kappa |\Delta(2\cos\vartheta_1, \ldots, 2\cos\vartheta_{\bar{N}})|^\beta \prod_{j=1}^{\bar{N}} \left(1 - \cos\vartheta_j\right)^a \left(1 + \cos\vartheta_j\right)^b \sin\vartheta_j \mathrm{d}\vartheta_j \tag{12.109}$$

on $[0, \pi]^{\bar{N}}$. The proportionality factor $\tilde{\zeta}_{\bar{N}}$ is defined in (12.48) and $\kappa = (\bar{N}-1)(-\frac{1}{2}\beta + a + b + 2)$. To achieve a pressure ramp of slope $-P/2\bar{N}$, one has to set

$$\beta = \frac{2P}{\bar{N}}, \quad a = b = -1 + \tfrac{1}{4}\beta. \tag{12.110}$$

Then, $\kappa = 0$ and under the measure

$$\prod_{j=0}^{N-2} (1 - \alpha_j^2)^{-1}(1 - \alpha_j^2)^{P(N-j-1)/N} \mathrm{d}\alpha_j \qquad (12.111)$$

the eigenvalues of C_N have the distribution

$$\zeta_{\bar{N}}(P)|\Delta(2\cos\vartheta_1, \ldots, 2\cos\vartheta_n)|^{2P/\bar{N}} \prod_{j=1}^{\bar{N}} (\sin\vartheta_j)^{((P/\bar{N})-1)} \mathrm{d}\vartheta_j. \qquad (12.112)$$

Now, the strategy of Section 12.3 is in force. In (12.112), we add a confining potential. The power of $\sin\vartheta_j$ converges to -1 and in the limit $\bar{N} \to \infty$, the free energy functional becomes

$$\mathcal{F}_{\mathrm{kv\triangleright}}^{\circ}(\varrho) = \int_0^{\pi} \mathrm{d}w\varrho(w) \left(V(w) + \log\sin w + \log\varrho(w) \right.$$

$$\left. - P \int_0^{\pi} \mathrm{d}w'\varrho(w') \log(2|\cos w - \cos w'|) \right). \qquad (12.113)$$

$\mathcal{F}_{\mathrm{kv\triangleright}}^{\circ}$ has to be minimized over all $\varrho \geqslant 0$ with $\int_0^{\pi} \mathrm{d}w\varrho(w) = 1$ and boundary condition $\rho(-1) = \rho(1)$. Actually, our mean-field limit is somewhat singular, since at $a = b = -1 + (P/N)$, the *a priori* measure is barely integrable. The quadratic energy term is repulsive, but the linear $\log\sin$ term pushes the eigenvalues toward the two end points. It is not so obvious, whether and how the two terms balance. Fortunately, the particular case $V = 0$ has been studied in detail, thereby confirming (12.113) and a normalizable DOS.

Surprisingly, deviating from the conventional script, the confining potential V is corrected by the $\log\sin$ potential, which is attractive and favors accumulation of eigenvalues at $0, \pi$. The prior computation of the average currents is carried out for fixed V, which now has to be replaced by $\tilde{V}(w) = V(w) + \log(\sin w)$. Of course, ρ_{n} and ρ_{p} depend on $\tilde{V}$. Also, the operator T has to be adjusted to

$$Tf(w) = 2\int_0^{\pi} \mathrm{d}w' \log|2(\cos w - \cos w')|f(w'), \qquad (12.114)$$

$w \in [0, \pi]$. With these modifications, the TBA formalism can be taken over with no further changes. In Eq. (6.19), the bare velocity v is replaced by $2\cos v$ and the kernel $2\log|v - w|$ is substituted by $2\log|2(\cos v - \cos w)|$.

Then, the effective velocity and the shift are given by

$$v^{\text{eff}} = \frac{(-2\varsigma_1)^{\text{dr}}}{\varsigma_0^{\text{dr}}}, \quad q_1 = -2\nu \int_0^\pi dw \rho_{\text{p}}(w) \cos w \qquad (12.115)$$

upon adopting the dressing operator T from (12.114) and the confining potential $\tilde{V}$ instead of V. The hydrodynamic equations are argued along the standard route.

The discrete modified KdV equation is an integrable wave equation with no obvious Hamiltonian structure. Compared to the Ablowitz–Ladik lattice, CUE is replaced by COE. Interestingly enough, in the standard coordinates, an additional confining potential is generated. Such a feature has not yet been encountered before.

Notes and references

Section 12.1

The most useful sources of information are the monographs by Ablowitz and Segur (1981) and by Ablowitz *et al.* (2004). In the book by Grébert and Kappeler (2014), the integrable structure of continuum NLS is discussed in great detail. Equilibrium measures were first studied by Lebowitz *et al.* (1988), see also the more recent contributions by Oh and Quastel (2013) and Fröhlich *et al.* (2017). More general nonlinearities are covered in Bourgain (1994). GGEs with V a finite polynomial are studied by Zhidkov (2001). The discretization discussed in this section is based on Ablowitz and Ladik (1975, 1976). The GHD of the continuum sinh-Gordon equation has been developed by Bastianello *et al.* (2018), see also De Luca and Mussardo (2016). The integrable lattice version is reported by Orfanidis (1978). The integrable structure of continuum spin chain has been studied first by Takhtajan (1977) and Sklyanin (1979). Its integrable discretization is covered by Faddeev and Takhtajan (2007), see also the more recent contribution by Krajnik *et al.* (2021). Numerical simulations of the spin chain are reported by Das *et al.* (2019). The article by Spohn and Lebowitz (1977) is an early investigation of what is now called GGE for wave equations.

Section 12.2

Nenciu (2005, 2006) discovered the role played by the Cantero–Moral-Velázquez matrices, which were originally introduced with distinct goals, see Cantero *et al.* (2003, 2005). Further accounts are Killip and Nenciu (2007) and Simon (2007).

Section 12.3

The change of volume elements for the linear pressure ramp has been accomplished by Killip and Nenciu (2004) with motivations from the corresponding result by Dumitriu and Edelman (2002) for the Toda lattice. The circular log gas is summarized in Forrester (2010) and its mean-field version is studied by Trinh and Trinh (2021) and Hardy and Lambert (2021). The results reported in the text are based on Spohn (2022). Independently, similar results are worked out by Mazzuca and Grava (2023). In particular, for polynomial confining potentials, the free energy is proved to be the solution of the variational problem in (12.59). Mazzuca and Memin (2022) extend such results to all continuous confining potentials V, which conceptually is an important step. More generally, one would like to prove the following property of GHD. One starts from an initial profile $x \mapsto \nu\rho_{\mathsf{p}}(x)$, such that every $\nu\rho_{\mathsf{p}}(x)$ determines a unique GGE. To be shown is propagation of such a property in time. Numerical simulations of the Ablowitz–Ladik chain are reported by Mendl and Spohn (2015) who confirm ballistic scaling and broad spectra. A quantitative comparison with GHD is still missing. Based on Killip and Nenciu (2007) the N particle scattering is elucidated in Brollo and Spohn (2023).

Section 12.5

From a general perspective, the expressions (12.91) and (12.115) are somewhat puzzling. The numerator is the dressing of the derivative of energy, i.e., $-\sin w$ for (12.91) and $\cos w$ for (12.115). However, according to the rules, the denominator should be the derivative of momentum, thus respectively $\cos w$ and $-\sin w$, while we obtained the dressing of the constant function.

Section 12.6

The notion Schur flow for the dynamics generated by (12.97) goes back to Golinskiĭ (2006). The transformation of measure is proved by Killip and Nenciu (2004). In Trinh and Trinh (2021) and Forrester and Mazzuca (2021), explicit formulas are obtained for the thermal case with $a, b > -1$. With the methods Trinh and Trinh (2021) also the diagonal limit, $a = b = -1 + (P/N)$, is proved to yield the stated result. Surprisingly, the two-fold degeneracy of the spectrum of the CMV matrix results in asymptotically two-particle bound states, see Brollo and Spohn (2023).

Chapter 13

Hydrodynamics for the Lieb–Liniger δ-Bose Gas

Much of the interest related to GHD is rooted in the study of integrable *quantum* many-body systems. Over the past decade, the number of contributions staggered and it would be difficult to present a balanced view. Reviews and journal special volumes are available. On the other hand, without mentioning quantum models, the GHD story would be utterly incomplete. Our compromise consists of two elements. The first acquisition is the quantized version of the Toda lattice, the topic of Chapter 14. Due to missing physical realizations, this model received less attention. Still, the model is instructive because an in-depth theoretical comparison between classical and quantum is feasible.

The most widely studied integrable quantum model by far is the quantized nonlinear Schrödinger equation, better known as the δ-Bose gas because the interaction between bosons is point-like. In their pioneering work from 1963, Elliott Lieb and Werner Liniger established that eigenvalues and eigenfunctions of the Hamiltonian can be obtained through the Bethe ansatz. This discovery triggered the field of integrable quantum many-body systems. Also, for GHD, the Lieb–Liniger model plays a very special role. It serves as paradigm for the study of more difficult models as the XXZ quantum spin chain and the spin-$\frac{1}{2}$ Fermi–Hubbard model. Even more importantly, rubidium atoms enclosed in a narrow tube are well modeled by bosons interacting through an extremely short-range potential. Experimental realizations are available through the amazing advances in cold atom physics, see the Insert in Section 13.3. Thus, as a second acquisition, we discuss the aspects of GHD for the δ-Bose gas. In particular, we will emphasize the structural similarity with regard to the DOS of a classical Lax matrix.

For all quantum mechanical models, the hydrodynamic scale is particle-based. The most convincing evidence is the computation of the two-particle scattering shift, which is obtained from the scattering of two quantum particles. Mathematically, this is very different from a two-soliton solution. The quantum particles are described by a wave equation in $\mathbb{R}^2$ governing the motion of the two-particle wave function $\psi(x_1, x_2, t)$, while a two-soliton wave field refers to a wave equation over $\mathbb{R}$. The construction of soliton-based hydrodynamics for quantum models remains as a task for the future.

13.1 Bethe ansatz

To quantize the continuum nonlinear Schrödinger equation of the introduction to Chapter 12, one starts from a scalar bosonic field $\Psi(x)$, $x \in \mathbb{R}$, with canonical commutation relations

$$[\Psi(x), \Psi(x')^*] = \delta(x - x'), \quad [\Psi(x), \Psi(x')] = 0, \quad [\Psi(x)^*, \Psi(x')^*] = 0.$$
(13.1)

In normal ordered form, the Lieb–Liniger Hamiltonian then reads

$$H_{\mathrm{li}} = \tfrac{1}{2} \int_{\mathbb{R}} \mathrm{d}x \big(\partial_x \Psi(x)^* \partial_x \Psi(x) + c \Psi(x)^* \Psi(x)^* \Psi(x) \Psi(x) \big).$$
(13.2)

Here, c is the coupling constant with $c \geqslant 0$ imposed. The attractive case, $c < 0$, is considerably more complicated because of bound states. In the widely accepted standard convention, the factor $\tfrac{1}{2}$ in front of the integral is omitted. We inserted here so as to have an energy–momentum relation of the free theory identical to the one for the noninteracting Toda fluid. The Lieb–Liniger Hamiltonian is formal because of point interactions. Fortunately, there is the first quantized version for which the Hamiltonian in the N-particle sector reads

$$H_{\mathrm{li},N} = -\sum_{j=1}^{N} \tfrac{1}{2}(\partial_{x_j})^2 + c \sum_{i,j=1, i<j}^{N} \delta(x_i - x_j).$$
(13.3)

Bosonic means that the operator acts on wave functions symmetric under particle exchange. The case of interest is a spatial interval $[0, \ell]$, $\ell > 0$, with periodic boundary conditions, which will be distinguished by either a sub- or superscript ℓ. Of course, one could also consider a finite number of particles moving on the real line, which would be the natural set-up for multi-particle scattering.

Let us first recall how to handle the delta potential. The simplest example is $N = 2$, for which the relative motion, $y = x_2 - x_1$, is governed by the Hamiltonian

$$H_{\mathrm{rel}} = -\partial_y^2 + c\,\delta(y). \qquad (13.4)$$

Away from $y = 0$, $H_{\mathrm{rel}}\psi(y) = -\partial_y^2\psi(y)$ with $\psi \in \mathcal{C}^2(\mathbb{R}\backslash\{0\})$, the twice continuously differentiable functions on $\mathbb{R}\backslash\{0\}$ with bounded derivatives. The δ-potential translates to the boundary condition

$$\partial_y\psi(0_+) - \partial_y\psi(0_-) = 2c\psi(0). \qquad (13.5)$$

The same mechanism works for the N-particle Hamiltonian $H_{\mathrm{li},\ell,N}$ acting on symmetric wave functions. For them, one can restrict the construction to the Weyl chamber $\mathbb{W}_{N,\ell} = \{0 \leqslant x_1 \leqslant \cdots \leqslant x_N \leqslant \ell\}$. By permutation symmetry, the Hamiltonian can then be extended to all other sectors. Away from the boundary, the Hamiltonian is the N-dimensional Laplacian,

$$H_{\mathrm{li},\ell,N} = -\sum_{j=1}^{N} \tfrac{1}{2}(\partial_{x_j})^2, \qquad (13.6)$$

which acts on $\psi \in \mathcal{C}^2(\mathbb{W}_{N,\ell} \backslash \partial\mathbb{W}_{N,\ell})$. Periodic boundary conditions mean

$$\psi(0, x^\perp) = \psi(x^\perp, \ell), \quad \partial_{x_1}\psi(0, x^\perp) = \partial_{x_N}\psi(x^\perp, \ell), \qquad (13.7)$$

while the interaction is encoded by the boundary condition

$$(\partial_{x_{j+1}} - \partial_{x_j})\psi(x) = c\psi(x)\big|_{x_j = x_{j+1}} \qquad (13.8)$$

with $j = 1, \ldots, N-1$, where the limit is taken from the interior of $\mathbb{W}_{N,\ell}$. It can be shown that the Laplacian with these boundary conditions defines a unique self-adjoint operator on $L^2(\mathbb{W}_{N,\ell})$.

On $\mathbb{W}_{N,\ell}$, an eigenfunction of $H_{\mathrm{li},\ell,N}$ has the Bethe ansatz form

$$\psi(x_1, \ldots, x_N) = \sum_{\sigma \in S_N} \left[\exp\left(\mathrm{i} \sum_{j=1}^{N} k_{\sigma(j)} x_j \right) \prod_{\substack{i<j \\ \sigma(i)>\sigma(j)}} A(k_i, k_j) \right], \qquad (13.9)$$

where the sum is over all permutations σ of $(1, \ldots, N)$. The k_js are called rapidities so as to distinguish from the momenta of noninteracting particles. ψ has to satisfy the boundary conditions (13.8), which leads to

$$A(k_1, k_2) = \frac{\mathrm{i}(k_1 - k_2) - c}{\mathrm{i}(k_1 - k_2) + c} = \mathrm{e}^{-\mathrm{i}\theta_{\mathrm{li}}(k_1 - k_2)}. \qquad (13.10)$$

Here, the two-particle phase shift equals

$$\theta_{\text{li}}(w) = 2\arctan(w/c), \tag{13.11}$$

implying the scattering shift

$$\theta'_{\text{li}}(w) = \frac{2c}{w^2 + c^2} = \phi_{\text{li}}(w). \tag{13.12}$$

Thus, when two bosons scatter, their distance is reduced in comparison to the free particle motion. Equivalently, one can think of a delay time. When two bosons scatter, they effectively stick for a short time span which depends on the incoming momenta.

Imposing periodic boundary conditions as in (13.7), the rapidities have to satisfy

$$e^{ik_j\ell} = (-1)^{N-1} \prod_{i=1}^{N} e^{-i\theta_{\text{li}}(k_j - k_i)}. \tag{13.13}$$

Equivalently, $k_1, \ldots, k_N$ solve the *Bethe equations*

$$2\pi I_j = \ell k_j + \sum_{i=1}^{N} \theta_{\text{li}}(k_j - k_i), \tag{13.14}$$

$j = 1, \ldots, N$, and hence are also referred to as *Bethe roots*. The quantum states are labeled by vectors $I = (I_1, \ldots, I_N)$, constrained as $I_1 < \cdots < I_N$ and with entries from $\mathbb{Z}$ in case of odd and from $\mathbb{Z} + \frac{1}{2}$ in case of even N. The set of quantum numbers is denoted by $\mathbb{I}_N$. For each such $I \in \mathbb{I}_N$, the Bethe equations have a unique solution denoted by $k(I) = k = (k_1, \ldots, k_N)$ ordered as $k_1 < \cdots < k_N$. It is known that these eigenfunctions constitute a complete orthonormal basis in $L^2(\mathbb{W}_{N,\ell})$.

◆◆ *Volume factors*: If the quantum numbers above would be redefined to take values in $2\pi\mathbb{Z}$, resp. $2\pi(\mathbb{Z} + \frac{1}{2})$, then the factors of 2π would disappear from our formulas. Thereby an optically even closer similarity to the Toda fluid would be achieved. For the ease of comparison, the conventional notation is adopted, however.   ◆◆

Symmetric eigenfunctions on $L^2([0, \ell]^N)$ can be written as

$$\psi_k(x_1, \ldots, x_N) = \sum_{\sigma \in S_N} (-1)^{\text{sgn}(\sigma)} \exp\left(i\sum_{j=1}^{N} k_{\sigma(j)} x_j\right)$$

$$\times \prod_{1 \leqslant i < j \leqslant N} (k_{\sigma(i)} - k_{\sigma(j)} + ic\,\text{sgn}(x_i - x_j)) \tag{13.15}$$

with $\mathrm{sgn}(\sigma)$ the sign of the permutation σ. ψ_k is not normalized, the normalized state vector being denoted by $|k\rangle = \langle \psi_k, \psi_k \rangle^{-\frac{1}{2}} \psi_k$. The normalization constants are known and the wave functions $|k\rangle$ span the bosonic subspace $L^2([0,\ell]^N)_{\mathrm{sym}}$. In particular,

$$H_{\mathrm{li},N} = \sum_{I \in \mathbb{I}_N} \left(\tfrac{1}{2} \sum_{j=1}^N (k_j)^2 \right) |k\rangle\langle k| \qquad (13.16)$$

with $k = k(I)$. Up to the prefactor $\frac{1}{2}$, if the power 2 is substituted by 0, one arrives at the number operator. Similarly, power 1 results in the total momentum. This suggests to introduce the nth conserved field ($=$ charge) by

$$Q^{[n],N} = \sum_{I \in \mathbb{I}_N} \left(\sum_{j=1}^N (k_j)^n \right) |k\rangle\langle k|. \qquad (13.17)$$

13.2 Bethe root densities, free energy, TBA equations

The Boltzmann weight involves a linear combination of charges. Therefore, again following the Toda lattice blueprint, we introduce the confining potential V and set

$$Q^{[V],N} = \sum_{I \in \mathbb{I}_N} \left(\sum_{j=1}^N V(k_j) \right) |k\rangle\langle k|. \qquad (13.18)$$

The precise class of confining potentials remains yet to be studied. A natural choice seems to be V continuous and $V(w) \geqslant c_0 + c_1|w|$ with $c_1 > 0$. The unnormalized GGE density matrix is defined by

$$\exp\left[-Q^{[V],N} \right] \qquad (13.19)$$

as an operator on $L^2([0,\ell]^N)_{\mathrm{sym}}$. Hence, the normalizing partition function is given by

$$Z_{\mathrm{li},N}(\ell, V) = \mathrm{tr}\left[e^{-Q^{[V],N}} \right] = \sum_{I \in \mathbb{I}_N} \exp\left[-\sum_{j=1}^N V(k_j) \right], \qquad (13.20)$$

trace over $L^2([0,\ell]^N)_{\mathrm{sym}}$. Correspondingly, the GGE average of some operator $\mathcal{O}$ is defined by

$$\langle \mathcal{O} \rangle_{\ell,V} = Z_{\mathrm{li},N}(\ell, V)^{-1} \mathrm{tr}\left[e^{-Q^{[V],N}} \mathcal{O} \right]. \qquad (13.21)$$

The Bethe roots depend on ℓ and, as before, this dependence is indicated in the partition function.

Our goal is the infinite volume

$$\ell \to \infty, \quad N = \nu\ell \tag{13.22}$$

with $\nu > 0$ and $\rho_f = 1/\nu$ the Bose gas density. As for previously studied models, the free energy of the δ-Bose gas will be determined through a variational problem. In fact, the free energy functional will be surprisingly similar to the one for the Toda and Calogero fluid, compare with Eqs. (9.47) and (11.58). Hence, it is of interest to have a closer look at the derivation. The second items are GGE averaged charges $\ell^{-1}\langle Q^{[n],N}\rangle_{\ell,V}$. These observables are diagonal in the energy basis, which reduces our task to an unconventional problem of classical statistical mechanics. As a simplification, the scattering shift appears already explicitly through the Bethe equations.

We first introduce the empirical *root density*

$$\varrho_N(w) = \frac{1}{N}\sum_{j=1}^{N}\delta(w - k_j), \tag{13.23}$$

compare with (3.25). Then,

$$\sum_{j=1}^{N}V(k_j) = N\int_{\mathbb{R}}dw\varrho_N(w)V(w) = N\mathcal{E}(\varrho_N), \tag{13.24}$$

an energy-like term, and our task is to express the sum over quantum states by a sum over root densities, at least approximately for large N. For given I, thus Bethe roots $k(I)$, we consider the equation

$$y(\lambda) = \nu\lambda + \theta_{li} * \varrho_N(\lambda) \tag{13.25}$$

with the convolution

$$\theta_{li} * \varrho_N(\lambda) = \int_{\mathbb{R}}dw\theta_{li}(\lambda - w)\varrho_N(w). \tag{13.26}$$

Since the function $y(\lambda)$ is strictly increasing, the relation

$$y(k_m^{vac}) = \frac{2\pi}{N}m, \tag{13.27}$$

$m \in \mathbb{Z}$, resp. $m \in \mathbb{Z}+\frac{1}{2}$, defines the increasingly ordered collection of points $\{k_m^{vac}\}$. This set is viewed as a collection of placeholders. Their empirical density is the empirical *space density*

$$\varrho_{s,N}(w) = \frac{1}{N}\sum_{m\in\mathbb{Z}}\delta(w - k_m^{vac}). \tag{13.28}$$

Since $y(k_j) = 2\pi I_j/N$, the Bethe roots are a subset of $\{k_m^{\text{vac}}\}$. These are particle locations in distorted momentum space. Particles partially fill the placeholder set $\{k_m^{\text{vac}}\}$. The complementary set, denoted by $\{k_m^{\text{h}}\}$, are hole locations in distorted momentum space. The empirical *hole density* is then

$$\varrho_{\text{h},N}(w) = \frac{1}{N} \sum_{m\in\mathbb{Z}} \delta(w - k_m^{\text{h}}). \tag{13.29}$$

Obviously,

$$\varrho_N + \varrho_{\text{h},N} = \varrho_{\text{s},N}. \tag{13.30}$$

We assume that, relative to the given GGE, with probability one, the empirical densities have a deterministic limit as $N \to \infty$. More precisely, one picks a well-localized test function, f, and asserts that

$$\lim_{N\to\infty} \frac{1}{N} \sum_{m\in\mathbb{Z}} f(k_m^{\text{vac}}) = \lim_{N\to\infty} \int_{\mathbb{R}} dw f(w)\varrho_{\text{s},N}(w) = \int_{\mathbb{R}} dw f(w)\varrho_{\text{s}}(w), \tag{13.31}$$

where the expression on the left-hand side is random, while its limit is non-random. Equation (13.31) defines the infinite volume space density ϱ_{s}, correspondingly for ϱ and ϱ_{h}. According to (13.30),

$$\varrho + \varrho_{\text{h}} = \varrho_{\text{s}}. \tag{13.32}$$

Next, we study the limit related to (13.25). Setting $\lambda = k_j^{\text{vac}}$ and summing over $f(k_j^{\text{vac}})$ yield

$$\frac{1}{N} \sum_{j\in\mathbb{Z}} f(k_j^{\text{vac}})\frac{2\pi}{N}j = \frac{1}{N} \sum_{j\in\mathbb{Z}} f(k_j^{\text{vac}})\left(\nu k_j^{\text{vac}} + \theta_{\text{li}} * \varrho_N(k_j^{\text{vac}})\right). \tag{13.33}$$

By construction, the right-hand side converges to

$$\int_{\mathbb{R}} dw f(w)\varrho_{\text{s}}(w)\left(\nu w + \theta_{\text{li}} * \varrho(w)\right). \tag{13.34}$$

For the left-hand side, setting $i < j$, we start from

$$\int_{k_i^{\text{vac}}}^{k_j^{\text{vac}}} dw \varrho_{\text{s},N}(w) = \frac{1}{N}(j - i). \tag{13.35}$$

For $i \to -\infty$, the nonlinearity becomes negligible and $k_i^{\text{vac}} = 2\pi i/N$. Therefore, by restricting the support of the test function to $[-a, \infty)$ and setting $k_i^{\text{vac}} = -a$ with a sufficiently large,

$$\frac{1}{N}\sum_{j\in\mathbb{Z}} f(k_j^{\text{vac}}) \left(\int_{-a}^{k_j^{\text{vac}}} 2\pi\varrho_{\text{s},N}(w')\mathrm{d}w' - a\right)$$

$$\to \int_{\mathbb{R}} \mathrm{d}w f(w)\varrho_{\text{s}}(w) \left(\int_{-a}^{w} 2\pi\varrho_{\text{s}}(w')\mathrm{d}w' - a\right) \qquad (13.36)$$

as $N \to \infty$. Inserting (13.34) and (13.36) in Eq. (13.33), one arrives at the pointwise identity

$$\int_{-a}^{w} 2\pi\varrho_{\text{s}}(w')\mathrm{d}w' - a = \nu w + \theta_{\text{li}} * \varrho(w), \qquad (13.37)$$

which upon differentiating yields

$$2\pi\varrho_{\text{s}}(w) = \nu + \phi_{\text{li}} * \varrho(w). \qquad (13.38)$$

Through the root density, the scattering shift rules the space density.

The final step would require more details. The energy term appeared already in (13.24). Hence, one still has to figure out the entropy, namely the number of Is subject to our constraints. For this purpose considered is a small volume element, $\mathrm{d}w$, which still contains a huge number of place-holders given by $N\varrho_{\text{s}}(w)\mathrm{d}w$. They split into Bethe roots, $N\varrho(w)\mathrm{d}w$, and holes, $N\varrho_{\text{h}}(w)\mathrm{d}w$. Merely counting the corresponding quantum numbers, the corresponding local entropy equals

$$\log \frac{(N\varrho_{\text{s}}(w)\mathrm{d}w)!}{(N\varrho(w)\mathrm{d}w)!(N\varrho_{\text{h}}(w)\mathrm{d}w)!}$$

$$\approx N\mathrm{d}w\big(\varrho_{\text{s}}(w)\log\varrho_{\text{s}}(w) - \varrho(w)\log\varrho(w) - \varrho_{\text{h}}(w)\log\varrho_{\text{h}}(w)\big), \quad (13.39)$$

where the linear term vanishes because of (13.32). With ϱ_{s} defined in (13.38) and ϱ_{h} in (13.32), the resulting entropy per particle is thus given by

$$\mathcal{S}(\varrho) = \int_{\mathbb{R}} \mathrm{d}w\big(\varrho_{\text{s}}\log\varrho_{\text{s}} - \varrho\log\varrho - \varrho_{\text{h}}\log\varrho_{\text{h}}\big). \qquad (13.40)$$

In approximation, one arrives at

$$Z_{\text{li},N}(\ell, V) \approx \int \mathcal{D}(\varrho)\mathrm{e}^{-N\big(\mathcal{E}(\varrho)-\mathcal{S}(\varrho)\big)} \qquad (13.41)$$

involving a somewhat vague sum over all root densities.

The exponent defines the Yang–Yang free energy functional

$$\mathcal{F}_{\text{yy}}^{\circ}(\varrho) = \nu^{-1}\int_{\mathbb{R}} \mathrm{d}w\big(\varrho V + \varrho\log\varrho + \varrho_{\text{h}}\log\varrho_{\text{h}} - \varrho_{\text{s}}\log\varrho_{\text{s}}\big). \qquad (13.42)$$

The free energy is per unit length, compare with (9.47) for the Toda fluid. It is convenient to switch to $\rho = \nu^{-1}\varrho$ and correspondingly $\rho_s = \nu^{-1}\varrho_s$, $\rho_h = \nu^{-1}\varrho_h$. We also introduce the T operator

$$Tf(w) = \frac{1}{2\pi} \int_{\mathbb{R}} dw' \phi_{li}(w - w') f(w'). \tag{13.43}$$

The kernel of $2\pi T$ equals the δ-Bose scattering shift. Also, $T\varsigma_0 = \varsigma_0$ and $T \to 1$ for $c \to 0$, while $T \to 0$ for $c \to \infty$. The latter case corresponds to free fermions and the former to free bosons. Eq. (13.32) becomes

$$\rho + \rho_h = \rho_s, \tag{13.44}$$

while (13.38) transforms to

$$2\pi\rho_s = 1 + 2\pi T\rho. \tag{13.45}$$

Then, using (13.44), the free energy per unit length is written as

$$\mathcal{F}_{yy}(\rho) = \int_{\mathbb{R}} dw \big(\rho V + \rho \log \rho + \rho_h \log \rho_h - (\rho + \rho_h) \log(\rho + \rho_h)\big). \tag{13.46}$$

The variation is over all ρ with $\rho \geqslant 0$, $\nu\langle\rho\rangle = 1$, and $\rho + \rho_h = (1/2\pi) + T\rho$. Note that, deviating from classical models, the factor 2π appears as a result of momentum quantization.

♦♦ *Convexity of the Yang–Yang free energy functional*: Following the original argument, we establish that the functional $\mathcal{F}_{yy}$ is convex. We start from two arbitrary spectral densities ρ_0, ρ_1 of well-defined Yang–Yang free energy and consider the linear interpolation $\rho_u = \rho_0(1-u) + u\rho_1$, $0 \leqslant u \leqslant 1$. The aim is to show that $\mathcal{F}_{yy}(\rho_u)'' > 0$, the prime denoting derivative with respect to u. Note that $\rho'_u = \rho_1 - \rho_0$ and, in general, $\rho' + \rho'_h = T\rho'$. With this input, differentiating $\mathcal{F}_{yy}(\rho(u))$ is straightforward with the result

$$\mathcal{F}_{yy}(\rho_u)'' = \int_{\mathbb{R}} dw \left(\frac{1}{\rho_u} + \frac{1}{\rho_{h,u}}\right) \left((\rho_1 - \rho_0) - \frac{\rho_u}{\rho_u + \rho_{h,u}} T(\rho_1 - \rho_0)\right)^2 \geqslant 0. \tag{13.47}$$

While likely to be correct, strict convexity would be a further step. ♦♦

We remove the constraint $\nu\langle\rho\rangle = 1$ by the Lagrange multiplier μ. Then, the free energy becomes

$$\mathcal{F}^{\bullet}_{yy}(\rho) = \int_{\mathbb{R}} dw \big(\rho(V - \mu) + \rho \log \rho + \rho_h \log \rho_h - (\rho + \rho_h) \log(\rho + \rho_h)\big). \tag{13.48}$$

The minimizing root density is the particle density ρ_p, which is the solution of the saddle-point equation

$$V - \mu - \log \frac{\rho_h}{\rho_p} - T \log \left(1 + \frac{\rho_p}{\rho_h}\right) = 0. \tag{13.49}$$

Here, it is understood that space and hole density associated with ρ_p are again denoted by ρ_s, ρ_h. Hence, $\rho_s = \rho_p + \rho_h$ and, in addition,

$$\rho_s = \frac{1}{2\pi} + T\rho_p, \tag{13.50}$$

which confirms the usage of "space density". For $c = 0$, i.e., $T = 0$, the Bethe roots are merely the eigenvalues of the quantized momentum, which in the infinite volume limit converge to the Lebesgue measure $\frac{1}{2\pi}dw$. Through the interaction, this density is modified, in the same spirit as the bare velocity is modified to v^{eff}. Particle and hole densities are then taken relative to such modified space density.

To solve (13.49) and (13.50), we introduce the number density

$$\rho_n = \frac{\rho_p}{\rho_p + \rho_h} = \frac{\rho_p}{\rho_s}, \tag{13.51}$$

thereby yielding the quasi-energy ε through

$$\rho_n = \frac{1}{1 + e^{\varepsilon}}, \quad e^{\varepsilon} = \frac{\rho_h}{\rho_p}. \tag{13.52}$$

Then, using the dressing transformation

$$f^{\text{dr}} = f + T\rho_n f^{\text{dr}}, \quad f^{\text{dr}} = \left(1 - T\rho_n\right)^{-1} f, \tag{13.53}$$

one obtains

$$2\pi\rho_s = \varsigma_0^{\text{dr}}. \tag{13.54}$$

With these conventions, Eq. (13.49) turns into the quantum TBA equation

$$\varepsilon = V - \mu - T \log(1 + e^{-\varepsilon}), \tag{13.55}$$

compare with (9.50) for the Toda fluid.

To determine the free energy of the δ-Bose gas, one has to evaluate the free energy functional $\mathcal{F}_{yy}^{\bullet}$ at its minimizer ρ_p, which leads to

$$\lim_{\ell \to \infty} -\frac{1}{\ell} \log Z_{\text{li},N}(\ell, \mu, V) = F_{yy}(\mu, V) = \mathcal{F}_{yy}^{\bullet}(\rho_p)$$

$$= -\frac{1}{2\pi} \int dw \log(1 + e^{-\varepsilon}), \tag{13.56}$$

a form familiar from noninteracting fermions. Adding the small perturbation $V + \kappa\varsigma_n$ and computing the linear response of the free energy in κ, one obtains

$$\lim_{\ell \to \infty} \ell^{-1} \langle Q^{[n],\ell} \rangle_{V,\ell} = \langle \rho_\mathsf{p} \varsigma_n \rangle \tag{13.57}$$

with $n = 0, 1 \dots$.

This identity suggests the proper interpretation of the minimizing particle density ρ_p. Under the GGE, the Bethe roots $k_1, \dots, k_N$ with variable N are random and combine into a empirical density as

$$\rho_{\mathsf{p},\ell}(w) = \ell^{-1} \sum_{j=1}^{N} \delta(w - k_j), \tag{13.58}$$

compare with Eq. (13.23). Obviously, $\rho_{\mathsf{p},\ell}(w) \geqslant 0$ with normalization $\langle \rho_{\mathsf{p},\ell} \rangle = \nu^{-1}$. Equation (13.57) states that on average the nth moment of $\rho_{\mathsf{p},\ell}$ converges to the one of ρ_p as $\ell \to \infty$. In Chapter 4, we explained that typical fluctuations of the empirical density of Lax matrix eigenvalues are of order $1/\sqrt{N}$. For the Lieb–Liniger model, one expects the same behavior and

$$\lim_{\ell \to \infty} \rho_{\mathsf{p},\ell}(w) = \rho_\mathsf{p}(w) \tag{13.59}$$

with probability one, upon integrating against a localized smooth test function.

13.3 Charge currents, hydrodynamic equations

To obtain the total currents, the usual strategy is to first figure out the charge densities. They satisfy a local conservation law from which one deduces the charge current density and thereby also the total charge currents. The total charges have been defined in (13.17). Then, to the lowest order and on the entire real line,

$$Q^{[0]}(x) = \Psi(x)^* \Psi(x), \quad Q^{[1]}(x) = -\mathrm{i}\Psi(x)^* \partial_x \Psi(x),$$
$$Q^{[2]}(x) = \partial_x \Psi(x)^* \partial_x \Psi(x) + c\Psi(x)^* \Psi(x)^* \Psi(x)\Psi(x), \tag{13.60}$$

which implies the current densities

$$J^{[0]}(x) = Q^{[1]}(x),$$
$$J^{[1]}(x) = \partial_x \Psi(x)^* \partial_x \Psi(x) + \tfrac{1}{2} c\Psi(x)^* \Psi(x)^* \Psi(x)\Psi(x). \tag{13.61}$$

Thus, a natural strategy would be to start from the classical nonlinear Schrödinger equation and its known sequence of charge densities involving higher-order field derivatives. As a naive guess, imposing normal order will yield the charges of the Lieb–Liniger model in the second quantized form, which would then be manifestly local. But this scheme works only up to $n = 3$. Already for $Q^{[4]}$, more complicated subtraction terms appear. For the total charge currents, similar difficulties will appear. We still freely use $Q^{[n]}(x)$ and $J^{[n]}(x)$, assuming that at some point these difficulties can be resolved. For infinite volume, formally, the charges satisfy a local conservation law of the form

$$\partial_t Q^{[n]}(x,t) + \partial_x J^{[n]}(x,t) = 0. \tag{13.62}$$

Just to be clear, here t refers to the unitary time evolution. For the operators at time zero, we omit the argument t, e.g., $J^{[n]}(x) = J^{[n]}(x, t = 0)$.

To compute the average current, one starts from a GGE at infinite volume, $\ell = \infty$, average being denoted as $\langle \cdot \rangle_V$ and truncation by $\langle fg \rangle_V^{\mathrm{c}} = \langle f \rangle_V \langle g \rangle_V$. Then, the charge–current correlator is given by

$$B_{m,n} = \int_{\mathbb{R}} \mathrm{d}x \langle J^{[m]}(x) Q^{[n]}(0) \rangle_V^{\mathrm{c}}. \tag{13.63}$$

While the definition looks asymmetric, in fact,

$$B_{m,n} = B_{n,m}, \tag{13.64}$$

as has been explained for other models before, compare with (6.8). Now, setting $\partial_0 = \partial_{\mu_0}$ and $\partial_1 = \partial_{\mu_1}$, one arrives at

$$\partial_0 \langle J^{[n]}(0) \rangle_V = -B_{n,0} = -B_{0,n}$$
$$= -\int_{\mathbb{R}} \mathrm{d}x \langle Q^{[1]}(x) Q^{[n]}(0) \rangle_V = \partial_1 \langle Q^{[n]}(0) \rangle_V = \partial_1 \langle \rho_{\mathsf{p}} \varsigma_n \rangle. \tag{13.65}$$

The ∂_0 derivative of the average current is related to a susceptibility, i.e., a particular second derivative of the generalized free energy. This is a doable problem and we explain a short path toward its solution.

Using the TBA equation (13.55) to differentiate ρ_{n} with respect to ∂_0 and ∂_1, one obtains the relations

$$\partial_0 \rho_{\mathsf{n}} = -\rho_{\mathsf{n}}(1 - \rho_{\mathsf{n}}) \varsigma_0^{\mathrm{dr}}, \quad \partial_1 \rho_{\mathsf{n}} = -\rho_{\mathsf{n}}(1 - \rho_{\mathsf{n}}) \varsigma_1^{\mathrm{dr}}. \tag{13.66}$$

Hence,

$$\varsigma_1^{\mathrm{dr}} \partial_0 \rho_{\mathsf{n}} = \varsigma_0^{\mathrm{dr}} \partial_1 \rho_{\mathsf{n}} \tag{13.67}$$

and, because $\partial_0 \partial_1 \rho_{\mathsf{n}} = \partial_1 \partial_0 \rho_{\mathsf{n}}$ for mixed derivatives,

$$\partial_1 \varsigma_0^{\mathrm{dr}} = \partial_0 \varsigma_1^{\mathrm{dr}}. \tag{13.68}$$

Setting

$$v^{\text{eff}} = \frac{\varsigma_1^{dr}}{\varsigma_0^{dr}} \qquad (13.69)$$

and using (13.67) and (13.68), one obtains

$$\partial_0(\rho_p v^{\text{eff}}) = \frac{1}{2\pi} \partial_0(\rho_n \varsigma_1^{dr}) = \frac{1}{2\pi} \left(\varsigma_1^{dr} \partial_0 \rho_n + \rho_n \partial_0 \varsigma_1^{dr} \right)$$
$$= \frac{1}{2\pi} \left(\varsigma_0^{dr} \partial_1 \rho_n + \rho_n \partial_1 \varsigma_0^{dr} \right) = \frac{1}{2\pi} \partial_1(\varsigma_0^{dr} \rho_n) = \partial_1 \rho_p. \qquad (13.70)$$

Altogether, we arrived at

$$\partial_0 \left(\langle J^{[n]}(0) \rangle_V - \langle \rho_p v^{\text{eff}} \varsigma_n \rangle \right) = 0. \qquad (13.71)$$

Thus, the round bracket is independent of μ_0. To determine the respective free parameter, one observes that in the limit $\mu_0 \to -\infty$ there are no particles. The GGE average current vanishes and so does ρ_p since $\nu \langle \rho_p \rangle = 1$. Hence, one concludes

$$\lim_{\ell \to \infty} \ell^{-1} \langle J^{[n],\ell} \rangle_{V,\ell} = \langle \rho_p v^{\text{eff}} \varsigma_n \rangle. \qquad (13.72)$$

As before, one confirms that the effective velocity can be written as a solution of the collision rate ansatz

$$v^{\text{eff}}(w) = w + \int_{\mathbb{R}} dw' \phi_{\text{li}}(w - w') \rho_p(w') \left(v^{\text{eff}}(w') - v^{\text{eff}}(w) \right). \qquad (13.73)$$

Assuming propagation of local GGE, one arrives at the hydrodynamic equation of the δ-Bose gas

$$\partial_t \rho_p(x, t; v) + \partial_x \left(v^{\text{eff}}(x, t; v) \rho_p(x, t; v) \right) = 0, \qquad (13.74)$$

which in structure is identical to the ones in the Toda and Calogero fluid. The transformation to an evolution equation for ρ_n remains valid with the expected result

$$\partial_t \rho_n(x, t; v) + v^{\text{eff}}(x, t; v) \partial_x \rho_n(x, t; v) = 0. \qquad (13.75)$$

◆◆ *Experimental realizations*: The 2006 pioneering experiment on the δ-Bose gas is widely known as quantum Newton's cradle. In a suitably engineered 2D optical lattice, ^{87}Rb atoms are trapped in an array of tubes, consisting of 1,000–8,000 tubes with approximately 40–250 atoms per tube. The motion of atoms in a single tube is well described by the Lieb–Liniger model, except for the additional harmonic trapping potential which in principle breaks integrability. One starts with a centered cloud of atoms and applies a Bragg pulse which yields the superposition of two states with a momentum shift $v_0 = \pm \hbar k$. Thus, the initial rapidity density is of the form

$$\tfrac{1}{2}\big(\rho_{\mathsf{p},\mathrm{therm}}(x,0;v-v_0) + \rho_{\mathsf{p},\mathrm{therm}}(x,0;v+v_0)\big), \qquad (13.76)$$

where "therm" indicates a thermal rapidity density at a fixed temperature. The cloud splits into left and right peaks, which start to oscillate due to the trapping potential and thus periodically interpenetrate each other. On longer time scales, the peaks dephase. If the dynamics of atoms would be nonintegrable, entirely different patterns would emerge. Thus, integrability of the Bose gas is confirmed, at least qualitatively. At the time, no theoretical frame was available. This has changed and good agreement with numerical simulations of GHD equations from (13.74) has been reported. However, Eq. (13.74) has to be modified so as to include a trapping potential,

$$\partial_t \rho_{\mathsf{p}}(x,t;v) + \partial_x \big(v^{\mathrm{eff}}(x,t;v)\rho_{\mathsf{p}}(x,t;w)\big) - \omega_0^2 x \partial_v \rho_{\mathsf{p}}(x,t;v) = 0 \quad (13.77)$$

with ω_0 the frequency of the trap potential.

A more recent experiment uses the atom on chip technique. Suitable electrical wiring on the chip generates a floating magnetic tube, considerably longer than a tube created by optical means. As a result, the experimental cloud consists of $4{,}600 \pm 100$ ^{87}Rb atoms. The initial state is a thermal cloud corresponding to an external potential V_{ex}. By additional wiring, a harmonic, $V_{\mathrm{ex}}(x) = x^2$, and double bump quartic, $V_{\mathrm{ex}}(x) = -x^2 + x^4$, can be realized. We omitted a scale factor, but the bump size is much smaller than the tube length. After the initial preparation, the cloud expands into the tube and the average density is recorded for several times. Density profiles are compared with numerical solutions of GHD and of conventional hydrodynamics (CHD), which is based on conservation of mass, momentum, and energy only. This would be the appropriate scheme for a nonintegrable 1D fluid. For the single bump initial condition, both simulations yield similar density profiles. Not so surprisingly, good agreement with the experimental data is recorded. For the double bump, the CHD solution develops steep gradients which eventually will lead to shock formation, while for GHD, the two bumps flatten out smoothly. Thus, CHD and GHD differ qualitatively. The experimental data agree well with GHD.

As one drawback of the atom on chip technique, the tube looses mass, roughly 25%, during a typical run. This effect is not included in (13.74). It took substantial efforts to understand how to incorporate to GHD such a seemingly innocent loss term.   ◆◆

13.4 Generic structure of TBA

While our results are somewhat scattered, we reached the stage to compare in more detail integrable classical and quantum fluids. For the Toda fluid, the free energy density is given by

$$F_{cl}(\varepsilon) = -e^{-\varepsilon}, \quad F_{cl}(\mu, V) = \int_{\mathbb{R}} dw\, F_{cl}(\varepsilon(w)) \tag{13.78}$$

and the number density is defined through

$$\rho_n = F_{cl}'(\varepsilon) = e^{-\varepsilon}. \tag{13.79}$$

The TBA equation reads

$$\varepsilon = V - \mu + T F_{cl}(\varepsilon). \tag{13.80}$$

Its solution is the number density ρ_n, which is linked to the first-order derivatives of the free energy as

$$\langle Q^{[n]}(0) \rangle_{\mu, V} = \langle \varsigma_0^{dr} \rho_n \varsigma_n \rangle = \langle \rho_p \varsigma_n \rangle. \tag{13.81}$$

The average currents are determined by the current potential with density

$$G_{cl}(w) = w F_{cl}(\varepsilon(w)), \quad G_{cl}(\mu, V) = \int_{\mathbb{R}} dw\, G_{cl}(w). \tag{13.82}$$

The first-order derivatives of G_{cl} are the average currents,

$$\langle J^{[n]}(0) \rangle_{\mu, V} = \langle \rho_p v^{eff} \varsigma_n \rangle, \quad v^{eff} = \varsigma_1^{dr} / \varsigma_0^{dr}. \tag{13.83}$$

The same formulas hold for the hard rod and Calogero fluids upon adjusting the two-particle scattering shift. In case of hard rod and Toda lattices, the free energy and the GGE averaged fields and currents are per lattice site. Therefore, the free energy has to be multiplied by the factor ν and the effective velocity has to be shifted by q_1. To be noted, the formulas above are specific for the single particle energy–momentum relation $E(v) = v^2/2$.

   For the Ablowitz–Ladik lattice, the w-integration is over the interval $[0, 2\pi]$ and the current potential density reads

$$G_{al}(w) = -2 \sin(w) F_{cl}(\varepsilon(w)), \tag{13.84}$$

which derives from $E(w) = 2 \cos w$ and hence $E'(w) = -2 \sin w$. The Schur flow has the anomaly of an additional self-generated confining potential.

Turning to the δ-Bose gas, except for the scattering shift, the *only* change is the free energy density, which now reads

$$F_{\mathrm{qu}}(\varepsilon) = -\log(1 + \mathrm{e}^{-\varepsilon}). \tag{13.85}$$

The free energy becomes

$$F_{\mathrm{qu}}(\mu, V) = \frac{1}{2\pi} \int_{\mathbb{R}} \mathrm{d}w F_{\mathrm{qu}}(\varepsilon(w)), \tag{13.86}$$

where the factor of $1/2\pi$ can be viewed as modifying the classical *a priori* measure $\mathrm{d}w$ to $\frac{1}{2\pi}\mathrm{d}w$. The number density results as

$$\rho_{\mathrm{n}} = F'_{\mathrm{qu}}(\varepsilon) = \frac{1}{1 + \mathrm{e}^{\varepsilon}}. \tag{13.87}$$

Finally, the current potential and its density become

$$G_{\mathrm{qu}}(w) = wF_{\mathrm{qu}}(\varepsilon(w)), \quad G_{\mathrm{qu}}(\mu, V) = \frac{1}{2\pi} \int_{\mathbb{R}} \mathrm{d}w G_{\mathrm{qu}}(\mu, V). \tag{13.88}$$

To be discussed in Chapter 14, the quantized Toda chain follows the same scheme.

A convincing pattern seems to emerge. For a given integrable model with particle-based hydrodynamics, one first has to identify the two-particle scattering shift and the thereby resulting space of spectral parameters. The distinction between classical and quantum is merely reflected through adopting either F_{cl} or F_{qu}. The one-particle dispersion law enters in the definition of the current potential.

13.5 Gaudin matrix

To determine the GGE averaged currents, an interesting entirely disjoint approach has been developed recently, for which the quantum mechanical structure of the Lieb–Liniger model enters in a more fundamental way. While we cannot write down so easily explicit formulas, abstractly there must exist the total current operator $J^{[n],\ell}$ for the total charge $Q^{[n],\ell}$ as defined in Eq. (13.17) with $\ell = \nu N$ and N the dimension of the vector k. For this operator, one conjectures the matrix elements $\langle k | J^{[n],\ell} | k \rangle$, which suffices for the computation of GGE expectations. The central object is the Gaudin matrix defined by

$$(G^{\ell})_{i,j} = \delta_{ij} \left(\ell + \sum_{m=1}^{N} 2\pi T(k_j, k_m) \right) - 2\pi T(k_i, k_j), \tag{13.89}$$

where $T(k_i, k_j)$ is the kernel of the integral operator defined in (13.43). Then, the claim is

$$\langle k | J^{[n],\ell} | k \rangle = \ell \sum_{i,j=1}^{N} \left((G^\ell)^{-1} \right)_{i,j} (k_i)^n k_j. \tag{13.90}$$

The prefactor ℓ arises because by stationarity the current density has to be independent of x.

Trusting in (13.90), the GGE averaged currents can be obtained. We fix a GGE with some confining potential V. Under this GGE, the Gaudin matrix is a random matrix. Presumably its large N, ℓ limit does not exist. But we have to invert the Gaudin matrix only on very special vectors. Therefore, we pick test functions f, g and consider the quadratic form

$$\ell^{-2} \sum_{i,j=1}^{N} f(k_i)(G^\ell)_{i,j} g(k_j) = \int_{\mathbb{R}^2} dw dw' f(w) \mathsf{G}_\ell(w, w') g(w'). \tag{13.91}$$

The left-hand side can be expressed as quadratic functional of the empirical density integrated against some smooth functions. Therefore, relying on (13.59), one obtains the limit

$$\lim_{\ell \to \infty} \langle f, \mathsf{G}_\ell g \rangle = \langle \rho_{\mathsf{p}} f, \mathsf{G} g \rangle \tag{13.92}$$

with probability one, where the limiting operator is defined through

$$\mathsf{G} g(w) = g(w) + 2\pi \int_{\mathbb{R}} dw' T(w, w') \rho_{\mathsf{p}}(w') \big(g(w') - g(w) \big). \tag{13.93}$$

Since the limit is non-random, unless badly behaved near zero, also the inverse converges to the inverse of the limit with probability one, i.e.,

$$\lim_{\ell \to \infty} \langle f, (\mathsf{G}_\ell)^{-1} g \rangle = \langle \rho_{\mathsf{p}} f, \mathsf{G}^{-1} g \rangle. \tag{13.94}$$

In our particular application, $f(w) = w^n$ and $g(w) = w$. Hence,

$$\lim_{\ell \to \infty} \ell^{-1} \langle J^{[n],\ell} \rangle_{V,\ell} = \langle \rho_{\mathsf{p}} \varsigma_n, \mathsf{G}^{-1} \varsigma_1 \rangle. \tag{13.95}$$

Defining $\mathsf{G}^{-1} \varsigma_1 = \bar{v}$, one obtains

$$\bar{v}(w) = w + 2\pi \big(T(\rho_{\mathsf{p}} \bar{v})(w) - T\rho_{\mathsf{p}}(w) \bar{v}(w) \big). \tag{13.96}$$

Hence, $\bar{v} = v^{\mathrm{eff}}$ and

$$\lim_{\ell \to \infty} \ell^{-1} \langle J^{[n],\ell} \rangle_{V,\ell} = \langle \rho_{\mathsf{p}} v^{\mathrm{eff}} \varsigma_n \rangle, \tag{13.97}$$

in agreement with (13.72).

♦♦ *Number and momentum current*: In second quantization, the particle current equals the momentum,

$$Q^{[1],\ell}(x) = -i\Psi(x)^*\partial_x\Psi(x), \quad Q^{[1],\ell} = -i\int_0^\ell dx\Psi(x)^*\partial_x\Psi(x) = J^{[0],\ell}.$$
(13.98)

Considering $n = 0$ in (13.90) and using $(G_\ell 1)_i = 1$ yield the result

$$\langle k|J^{[0],\ell}|k\rangle = \ell\sum_{i,j=1}^N ((G^\ell)^{-1})_{i,j}(k_i)^0 k_j = \sum_{j=1}^N k_j,$$
(13.99)

which agrees with (13.98).

To determine the momentum current, the left-hand part of (13.98) is integrated over some test function g. Working out the time derivative yields

$$i\left[H_{\mathrm{li},\ell}, -i\int_0^\ell dxg(x)\Psi(x)^*\partial_x\Psi(x)\right] = \tfrac{1}{2}\int_0^\ell dx\Big((g''(x)\Psi(x)^*$$

$$+ 2g'(x)\partial_x\Psi(x)^*)\partial_x\Psi(x) + g'(x)c\Psi(x)^*\Psi(x)^*\Psi(x)\Psi(x)\Big).$$
(13.100)

By the conservation law, the total current is obtained by setting $g' = 1$,

$$J^{[1],\ell} = \int_0^\ell dx\big(\partial_x\Psi(x)^*\partial_x\Psi(x) + \tfrac{1}{2}c\Psi(x)^*\Psi(x)^*\Psi(x)\Psi(x)\big)$$

$$= 2H_{\mathrm{li},\ell} - \tfrac{1}{2}c\int_0^\ell dx\Psi(x)^*\Psi(x)^*\Psi(x)\Psi(x).$$
(13.101)

Now taking expectations with respect to $|k\rangle$ with $N = 2$ on both sides yields

$$\langle k|J^{[1],\ell}|k\rangle = 2\langle k|H_{\mathrm{li},\ell}|k\rangle - \tfrac{1}{2}c\int_0^\ell dx\langle k|\Psi(x)^*\Psi(x)^*\Psi(x)\Psi(x)|k\rangle$$

$$= k_1^2 + k_2^2 - \frac{1}{\ell + 2T(k_1, k_2)}\frac{2c(k_1 - k_2)^2}{(k_1 - k_2)^2 + c^2}.$$
(13.102)

Since the Gaudin matrix is a 2×2 matrix, (13.102) agrees with (13.90) for $N = 2, n = 1$.

To proceed to $N > 2$, such a brute force computation no longer works, which is a generic experience in this area. Instead, we recall that the momentum current equals the pressure P, and for energy $E(\ell)$, one has $P = -dE/d\ell$. These thermodynamic relations extend to a single eigenstate of the Lieb–Liniger model. We start from

$$E(\ell) = \tfrac{1}{2}\sum_{j=1}^{N} k_j^2, \quad -\frac{\mathrm{d}}{\mathrm{d}\ell}E(\ell) = \sum_{j,k=1}^{N} ((G_{N,\ell})^{-1})_{j,k}k_j k_k, \quad (13.103)$$

where the second identity is obtained by differentiating Eq. (13.14) with respect to ℓ. Let us consider a volume of length ℓ and allow in the Lieb–Liniger Hamiltonian a kinetic energy of strength κ, as $H_{\mathrm{li},\ell,\kappa,c} = \kappa T_{\mathrm{kin},\ell} + c T_{\mathrm{pot},\ell}$. By continuity, one can follow one particular vector of Bethe roots, k, at fixed N in their dependence on ℓ, κ, c. By spatial dilation, the corresponding energy then scales as $E(\ell, \kappa, c) = E(\alpha\ell, \alpha^{-2}\kappa, \alpha^{-1}c)$, where α^{-2} results from the Laplacian and α^{-1} from the first-order derivative of the boundary condition in (13.8). To first order in the deviation from ℓ, there is an overall factor of $-\ell^{-1}$ and a factor 2 for the kinetic energy and a factor 1 for the potential energy, in agreement with (13.16). Hence,

$$\frac{\mathrm{d}}{\mathrm{d}\ell}\langle k|H_{\mathrm{li},\ell,N}|k\rangle = -\ell^{-1}\langle k|J^{[1],\ell}|k\rangle. \quad (13.104)$$

Combining (13.103) and (13.104), we conclude that Eq. (13.90) specialized to $n = 1$ holds for arbitrary N.

It is remarkable, when averaging over a single energy eigenstate, one finds already a structural dependence which persists in the limit of large ℓ, N. No such features seem to be known for the Toda lattice or other classical integrable many-particle systems. ◆◆

Notes and references

Section 13.0

Transport in one-dimensional quantum lattice models, including integrable chains, is reviewed in Bertini *et al.* (2021). Conformal field theories out of equilibrium are investigated by Bernard and Doyon (2016). Generalized hydrodynamics for the XXZ chain is studied by Bertini *et al.* (2016) and Bulchandani *et al.* (2017) and the spin-$\frac{1}{2}$ Fermi–Hubbard model by Ilievski and De Nardis (2017a). The lecture notes of Doyon (2020) provide a much more extensive list of references, see also the recent special volumes, one edited by Bastianello *et al.* (2022) and a second one by Abanov *et al.* (2022). Up to 1996, the exciting progress on integrable quantum many-body systems is well covered in Korepin *et al.* (1997).

Section 13.1

Elliott H. Lieb, together with Werner Liniger, discovered the Bethe ansatz for the δ-Bose gas and analyzed ground state properties in Lieb and Liniger (1963) and Lieb (1963). For the repulsive case, the completeness of Bethe wave functions has been proved by Dorlas (1993). The attractive δ-Bose gas is more complicated because of bound states. An exhaustive discussion of the structure of eigenfunctions can be found in Dotsenko (2010). In the attractive case, completeness is established in the thesis of Oxford (1979), see also the contribution by Prolhac and Spohn (2011). The local structure of higher charges is studied in Davies (1990) and Davies and Korepin (2011). For computations, the algebraic Bethe ansatz is often more powerful. An example is the norm of eigenfunctions, as discussed in Piroli and Calabrese (2015). The lecture notes by Franchini (2017) are a most readable introduction.

Section 13.2

The variational-type solution for the thermodynamics has been obtained by Yang and Yang (1969). A proof is accomplished by Dorlas *et al.* (1993) using methods from large deviations. The Bethe ansatz is very well covered in the literature, for example, see Takahashi (1999), Faddeev (1996), Sutherland (2004), and Gaudin (2014). At the time, the conventional thermodynamics was in focus. The GGE is discussed by Mossel and Caux (2012). It would be interesting to find out whether and how the methods in Dorlas *et al.* (1993) extend to a more general class of confining potentials. There have been several attempts in improving the Yang–Yang method. A recent study is Koslov *et al.* (2018) with references to earlier work.

Section 13.3

The GHD of the δ-Bose gas is discussed by Doyon (2020), see also Bonnemain *et al.* (2022). For the collision rate ansatz, we follow the contribution of Yoshimura and Spohn (2020).

Highly recommended reading is the comprehensive review by Bouchoule and Dubail (2021), see also Malvania *et al.* (2021). In addition to a theoretical section, the article discusses at length numerical schemes for solving GHD of the δ-Bose gas. The second half of the review is concerned with experimental realizations. The original quantum Newton cradle experiment was invented by Kinoshita *et al.* (2006), see also the more recent version Li *et al.* (2020). The GHD analysis of the experiment can be found in Caux *et al.* (2019). The atoms on chip techniques are described by Schemmer *et al.* (2019). The GHD handling of mass loss is accomplished in Bouchoule *et al.* (2020). The distinction between GHD and CHD is studied by Doyon *et al.* (2017). The connection of the Lieb–Liniger model to cold atom physics is reviewed by Zwerger (2022).

Section 13.4

Our discussion is based on the lecture notes by Doyon (2020). He points out three additional classes referring to classical phonons, classical radiation, and quantum bosons.

Section 13.5

For the discussion of the Gaudin matrix and its relation to currents, we follow the work of Borsi *et al.* (2020). Other methods rely on long range deformations, see Pozsgay (2020a), and properties of the boost operator, see Yoshimura and Spohn (2020). Corresponding results for spin systems are obtained by Pozsgay (2020).

Chapter 14

Quantum Toda Lattice

The classical Toda Hamiltonian is quantized according to standard rules. Thereby one obtains a system of N particles with Hilbert space $\mathcal{H}_N = L^2(\mathbb{R})^{\otimes N} = L^2(\mathbb{R}^N)$. The particles have position x_j and momenta p_j, $j = 1, \ldots, N$, satisfying the commutation relations $[p_i, x_j] = -\mathrm{i}\hbar\delta_{ij}$. In position space representation, the complex-valued wave functions are of the form $\psi(x_1, \ldots, x_N)$. For the jth particle, the position operator is multiplication by x_j and the momentum operator $p_j = -\mathrm{i}\hbar\partial_{x_j}$. In conventional notation, the interaction term reads

$$X_j = \mathrm{e}^{x_j - x_{j+1}}, \quad X_N = \mathrm{e}^{x_N - x_1}, \tag{14.1}$$

$j = 1, \ldots, N - 1$, where the latter condition imposes periodic boundary conditions. Then, the Hamiltonian of the quantum Toda lattice reads

$$H_{\mathrm{qt},N} = \sum_{j=1}^{N} \left(\tfrac{1}{2}p_j^2 + X_j\right). \tag{14.2}$$

Planck's constant $\hbar$ regulates the relative strength of kinetic and potential energy. In the same spirit, for the classical Toda chain, we could have introduced a mass parameter, which however can be absorbed through an appropriate rescaling of spacetime. The classical chain has no free model parameter. However, quantum mechanically $\hbar$ cannot be scaled and must be maintained as a relevant parameter. The semi-classical limit corresponds to $\hbar \to 0$. Another common choice is to explicitly introduce an interaction strength, η, and the inverse decay length of the potential through $\eta\,\mathrm{e}^{-\gamma x}$. Then, the standard form (14.2) is recovered upon replacing $\hbar^2$ by $\hbar^2\gamma^2/\eta$.

For the harmonic lattice, i.e., $X_j = \tfrac{1}{2}\omega_0^2(x_j - x_{j+1})^2$, one usually introduces creation and annihilation operators, satisfying $[a_i, a_j^*] = \delta_{ij}$.

The Hamiltonian is then quadratic in $\{a_i, a_j^*, i, j = 1, \ldots, N\}$ and describes bosonic excitations of the ground state. Formally, such a transformation can be implemented also for the Toda lattice, leading to a nonlinear interaction between bosons. For our purposes, this representation does not seem to be so useful.

Our strategy is to follow the trail laid out by the classical model. With some confidence, we will arrive at the appropriate hydrodynamic Euler equations. As might have been anticipated, there is an evident similarity with the Lieb–Liniger model.

14.1 Integrability, monodromy matrix

While there are examples for which integrability is maintained under quantization, no guarantee can be issued. Thus, our first task is to establish N local conservation laws. The elegant approach relies on the monodromy matrix, which is familiar from other integrable many-body systems. Let us start with the classical chain governed by the Hamiltonian (14.2) and its associated Lax matrix L_N, see Eq. (2.12). The characteristic polynomial of L_N is given by

$$\det(\lambda - L_N) = \lambda^N \left(\sum_{m=0}^{N-1} (-\lambda)^{-m} I_m + (-1)^N (\lambda)^{-N} (I_N - 2) \right), \quad (14.3)$$

which has the leading term $I_0 = 1$. The coefficients I_m are the *Hénon invariants* which are in involution, i.e., $\{I_m, I_n\} = 0$. Since the Hamiltonian is proportional to $I_1^2 - I_2$, the Hénon invariants are in fact conserved. To work out the details of such an expansion, a different expression for the determinant is convenient. We first define the 2×2 matrix

$$L_j(\lambda) = \begin{pmatrix} \lambda - p_j & e^{q_j} \\ -e^{-q_j} & 0 \end{pmatrix} \quad (14.4)$$

and, as the product of 2×2 matrices, the *monodromy matrix*

$$T_N(\lambda) = L_1(\lambda) \ldots L_N(\lambda). \quad (14.5)$$

The parameter λ is the *spectral parameter*. To emphasize the distinction, sometimes L_N is called the big Lax matrix and $L_j(\lambda)$ the little Lax matrix. While L_N is tridiagonal, the little Lax matrix operates in $\mathbb{C}^2$. However, the trace yields again the characteristic polynomial

$$\text{tr}[T_N(\lambda)] = \det(\lambda - L_N) + 2(-1)^N. \quad (14.6)$$

For the expansion in λ, one merely has to successively differentiate the matrix product in (14.5), which yields

$$I_m = \sum_{\{k+2\ell=m\}} p_{i_1} p_{i_2} \cdots p_{i_k} (-X_{j_1}) \cdots (-X_{j_\ell}), \qquad (14.7)$$

where the sum is over all strings of indices $i_1, i_2, \ldots, i_k, j_1, j_1 + 1, \ldots, j_\ell,$ $j_\ell + 1$. The indices are distinct and satisfy the constraint $m = k + 2\ell$, $k \geqslant 0, \ell \geqslant 0$. Two summands differing in the order of indices are counted only once. More visually, one considers a ring of N sites and has available k singletons and ℓ nearest neighbor pairs, the dominos. The ring is partially covered by singletons and dominos with no overlap allowed, which then defines the string of indices in (14.7). As an example, for $N = 4$, one obtains

$$I_1 = p_1 + p_2 + p_3 + p_4,$$

$$I_2 = p_1 p_2 + p_1 p_3 + p_1 p_4 + p_2 p_3 + p_2 p_4 + p_3 p_4 - X_1 - X_2 - X_3 - X_4,$$

$$I_3 = p_1 p_2 p_3 + p_1 p_2 p_4 + p_1 p_3 p_4 + p_2 p_3 p_4$$
$$\qquad - p_1 (X_2 + X_3) - p_2 (X_3 + X_4) - p_3 (X_4 + X_1) - p_4 (X_1 + X_2),$$

$$I_4 = p_1 p_2 p_3 p_4 - p_3 p_4 X_1 - p_1 p_4 X_2 - p_1 p_2 X_3 - p_2 p_3 X_4 + X_1 X_3 + X_2 X_4.$$

$$(14.8)$$

Except for $m = 1$, the I_ms are not local and have no meaningful density in the limit $N \to \infty$. To convert the Hénon invariants into a local form, we use the identity

$$\log \det (1 - \lambda^{-1} L_N) = - \sum_{m=1}^{\infty} \frac{1}{m} \lambda^{-m} \mathrm{tr}[(L_N)^m] = - \sum_{m=1}^{\infty} \frac{1}{m} \lambda^{-m} Q^{[m],N}.$$

$$(14.9)$$

As we know already from Section 2.1, the $Q^{[m],N}$s have indeed a local density.

In the quantum setting, $[p_j, X_m] = 0$, except for $j = m, m + 1$. Such terms do not appear in the sum (14.7) and no ambiguity in the operator ordering arises for the quantized version of the I_ms. This property strongly suggests that they will continue to be conserved quantities quantum mechanically.

Switching to the *quantum* Toda chain, there seems to be no analog of the Lax pair equation (2.14). However, as first observed by Sklyanin in 1985, the definition of monodromy matrix stays intact. As common usage, we switch from λ to u as a spectral parameter. The entries of the matrix $L_j(u)$ now become operators. As a consequence, the monodromy matrix $T_N(u)$, still defined as in (14.5), is a 2×2 matrix with operator entries. We introduce the R-matrix acting on $\mathbb{C}^2 \otimes \mathbb{C}^2$ by

$$R(u) = u\mathbf{1} \otimes \mathbf{1} - i\hbar P, \tag{14.10}$$

where P permutes the indices 1 and 2, i.e., $P\varphi_1 \otimes \varphi_2 = \varphi_2 \otimes \varphi_1$. Noting the commutation relations

$$[p_j, X_j] = -i\hbar X_j, \quad [p_{j+1}, X_j] = i\hbar X_j, \tag{14.11}$$

one confirms that

$$R(u - v)(L_j(u) \otimes \mathbf{1})(\mathbf{1} \otimes L_j(v)) = (\mathbf{1} \otimes L_j(v))(L_j(u) \otimes \mathbf{1})R(u - v) \tag{14.12}$$

and therefore also for the product

$$R(u - v)(T_N(u) \otimes \mathbf{1})(\mathbf{1} \otimes T_N(v)) = (\mathbf{1} \otimes T_N(v))(T_N(u) \otimes \mathbf{1})R(u - v). \tag{14.13}$$

Setting $\hat{t}_N(u) = \mathrm{tr}[T_N(u)]$, the trace of a 2×2-matrix, one obtains the commutation relations

$$\hat{t}_N(u)\hat{t}_N(v) = \hat{t}_N(v)\hat{t}_N(u) \tag{14.14}$$

valid for arbitrary u, v. $\hat{t}_N(u)$ is a polynomial of degree N, with operator-valued coefficients, the operators acting on $\mathcal{H}_N$. On abstract grounds, the relation (14.14) ensures that the Taylor coefficients of $\hat{t}_N(u)$ also commute with each other.

In fact, the family of operators can be worked out more concretely. Let us define

$$F_N(u) = \sum_{\{k+2\ell=N\}} (u - p_{i_1})(u - p_{i_2})\ldots(u - p_{i_k})(-X_{j_1})\ldots(-X_{j_\ell}), \tag{14.15}$$

where the string of indices satisfies the conditions listed below (14.7). We split as

$$F_N(u) = F_N^{\diamond}(u) + F_N^{\diamond\diamond}(u), \tag{14.16}$$

where $F_N^{\diamond\diamond}(u)$ collects all summands containing the factor e^{x_N}. Then, by an induction argument,

$$T_N(u) = \begin{pmatrix} F_N^{\diamond}(u) & e^{x_N}F_{N-1}^{\diamond}(u) \\ e^{-x_{N+1}}F_{N+1}^{\diamond\diamond}(u) & F_N^{\diamond\diamond}(u) \end{pmatrix}. \tag{14.17}$$

In particular,

$$\hat{t}_N(u) = F_N(u) = u^N \sum_{m=0}^{N} (-1)^m u^{-m} \hat{I}_m, \tag{14.18}$$

where $\hat{I}_m$ is the quantization of I_m. The $\hat{I}_m$-operators are non-local. To switch to local fields, we follow the recipe from the classical scheme. Since $\{\hat{I}_1, \ldots, \hat{I}_N\}$ is a family of commuting operators, one can still expand $\log(u^{-N}\hat{t}_N(u))$ at $u = \infty$ as

$$-\log\left(u^{-N}\hat{t}(u)\right) = u^{-1}\hat{Q}^{[1],N} + \tfrac{1}{2}u^{-2}\hat{Q}^{[2],N} \cdots$$
$$+ \tfrac{1}{N}u^{-N}\hat{Q}^{[N],N} + \mathcal{O}\left(u^{-(N+1)}\right), \tag{14.19}$$

which defines the quantized Flaschka invariants. By construction, $\hat{Q}^{[m],N}$ is a polynomial in $\hat{I}_1, \ldots, \hat{I}_m$ of maximal order m, which coincides with the corresponding classical identity. Thus, $\{\hat{Q}^{[1],N}, \ldots, \hat{Q}^{[N],N}\}$ is a family of commuting operators, also commuting with the $\hat{I}_1, \ldots, \hat{I}_N$. For the Hamiltonian,

$$H_{\mathrm{qt},N} = \tfrac{1}{2}\hat{Q}^{[2],N} = \tfrac{1}{2}\hat{I}_1^2 - \hat{I}_2, \tag{14.20}$$

which implies that the $\hat{I}_m$s and $\hat{Q}^{[m],N}$s are conserved.

For the particular case $N = 4$, the result is

$$\hat{Q}^{[1],4} = \hat{I}_1, \quad \hat{Q}^{[2],4} = \hat{I}_1^2 - 2\hat{I}_2, \quad \hat{Q}^{[3],4} = \hat{I}_1^3 - 3\hat{I}_1\hat{I}_2 + 3\hat{I}_3,$$
$$\hat{Q}^{[4],4} = \hat{I}_1^4 - 4\hat{I}_1^2\hat{I}_2 + 2\hat{I}_2^2 + 4\hat{I}_1\hat{I}_3 - 4\hat{I}_4. \tag{14.21}$$

Expanding the various powers, the corresponding charge densities are

$$\hat{Q}_j^{[1]} = p_j, \quad \hat{Q}_j^{[2]} = p_j^2 + 2X_j, \quad \hat{Q}_j^{[3]} = p_j^3 + 3(p_j + p_{j+1})X_j,$$
$$\hat{Q}_j^{[4]} = p_j^4 + 2(p_j^2 + p_j p_{j+1} + p_{j+1}^2)X_j$$
$$+ 2X_j(p_j^2 + p_j p_{j+1} + p_{j+1}^2) + 2X_j^2 + 4X_j X_{j+1}, \tag{14.22}$$

which are local, as anticipated. In fact, obtained is the minimal version of these densities, compare with the discussion above (2.28). The operator ordering is in force and relevant starting from $\hat{Q}^{[4],N}$ onward.

More generally, one can argue that $\hat{Q}^{[m],N}$ must be a particular operator ordering of $Q^{[m],N}$. If so, locality would follow from the fact that $Q^{[m],N}$ has a local density. Actually, the precise functional form of the higher order charges is fairly irrelevant for our purposes. As crucial information, the density of the mth charge depends on local operators in a block of size at most $(m/2)+1$. Since the Hamiltonian is the nearest neighbor, this property implies that the mth current density operator depends on a block of size at most $(m/2) + 2$.

14.2 Spectral properties

Since the interaction depends only on relative positions, an eigenfunction of $H_{\mathrm{qt},N}$ must be of the form

$$\psi_E(x_1,\dots,x_N) = \tilde{\psi}_{E_\perp}(x_2 - x_1,\dots,x_N - x_{N-1})\mathrm{e}^{\mathrm{i}\hbar^{-1}E_1 N^{-1}(x_1+\cdots+x_N)},$$

$$(14.23)$$

where $E = (E_1,\dots,E_N) = (E_1, E_\perp)$, implying $\hat{I}_1\psi_E = E_1\psi_E$. More abstractly, one can think of a fiber decomposition of $H_{\mathrm{qt},N}$ with respect to the total momentum. E_1 would then be the corresponding fiber parameter. In terms of the Hamiltonian, one explicitly splits into kinetic energy of the center of mass plus relative internal motion,

$$H_{\mathrm{qt},N} = H_{\mathrm{cm}} + H_{\mathrm{rel},N-1}. \qquad (14.24)$$

Thus, $\tilde{\psi}_{E_\perp} \in L^2(\mathbb{R})^{\otimes N-1}$ is an eigenfunction of $H_{\mathrm{rel},N-1}$. In fact, working out $H_{\mathrm{rel},N-1}$, one finds an interaction potential increasing exponentially in all directions. The eigenvalues of $H_{\mathrm{rel},N-1}$ are isolated, even non-degenerate, and the thermal operator $\exp(-H_{\mathrm{rel},N-1})$ is trace class.

It is of great advantage to consider simultaneously all eigenvalues of the Hénon invariants, to say

$$\hat{I}_m\psi_E = E_m\psi_E, \qquad (14.25)$$

$m = 1,\dots,N$, with E_1 as in (14.23). There is an explicit formula for the joint eigenfunctions and it is known that they are complete in the sense that linear combinations of the set $\{\tilde{\psi}_{E_\perp}\}$ span the entire Hilbert space $L^2(\mathbb{R}^{N-1})$. Then, ψ_E is also an eigenfunction of the transfer matrix $\hat{t}(u)$ with eigenvalue $\tau_N(u, E)$,

$$\hat{t}_N(u)\psi_E = \tau_N(u, E)\psi_E, \qquad (14.26)$$

where

$$\tau_N(u, E) = u^N\left(1 + \sum_{m=1}^{N}(-1)^m u^{-m} E_m\right). \qquad (14.27)$$

The mth Flaschka invariant has eigenvalue q_m,

$$\hat{Q}^{[m],N}\psi_E = q_m\psi_E. \qquad (14.28)$$

Since the operators commute, the q_ms are determined by

$$-\log\left(u^{-N}\tau_N(u, E)\right) = u^{-1}q_1 + \tfrac{1}{2}u^{-2}q_2 + \cdots + \tfrac{1}{N}u^{-N}q_N + \mathcal{O}\left(u^{-(N+1)}\right).$$

$$(14.29)$$

We now write

$$\tau_N(u) = \prod_{j=1}^{N}(u - \tau_j). \tag{14.30}$$

Then,

$$-\log\left(u^{-N}\tau_N(u)\right) = -\sum_{j=1}^{N}\log\left(1 - u^{-1}\tau_j\right)$$

$$= \sum_{m=1}^{N} u^{-m} \sum_{j=1}^{N} \tfrac{1}{m}(\tau_j)^m + \mathcal{O}\left(u^{-(N+1)}\right) \tag{14.31}$$

and

$$q_m = \sum_{j=1}^{N}(\tau_j)^m, \tag{14.32}$$

$m = 1, \ldots, N$. As already familiar, thereby one introduces an empirical density through

$$\varrho_N(w) = \frac{1}{N}\sum_{j=1}^{N}\delta(\tau_j - w) \tag{14.33}$$

with the property that

$$N^{-1}q_m = \int_{\mathbb{R}} dw \varrho_N(w) w^m. \tag{14.34}$$

Before continuing, we have to recall the quantum mechanical analog of the scattering shift studied in Section 2.3.2. The relative motion of two Toda particles is governed by the Schrödinger equation

$$i\partial_t \psi_t(x) = (-\hbar^2 \partial_x^2 + e^{-x})\psi_t(x). \tag{14.35}$$

The particle representing the relative motion travels inward from the far right with momentum $\hbar k_{\mathrm{in}}$, gets reflected at the potential barrier, and moves outward with momentum $\hbar k_{\mathrm{out}} = -\hbar k_{\mathrm{in}}$. The phase shift accumulated during the scattering process is given by

$$\theta_{\mathrm{qt}}(k) = k\log\hbar^2 - i\log\frac{\Gamma(1 + ik)}{\Gamma(1 - ik)}. \tag{14.36}$$

Surprisingly, the phase shift is independent of $\hbar$, except for the additive constant $k\log\hbar^2$. The scattering shift is then

$$\theta'_{\mathrm{qt}}(k) = \log\hbar^2 + \psi_{\mathrm{di}}(1 + ik) + \psi_{\mathrm{di}}(1 - ik) = \phi_{\mathrm{qt}}(k) \tag{14.37}$$

with Digamma function $\psi_{di} = \Gamma'/\Gamma$. From the convergent power series,

$$\psi_{di}(1 + ik) + \psi_{di}(1 - ik) = -\gamma_E + \sum_{n=1}^{\infty} \frac{k^2}{n(n^2 + k^2)}, \qquad (14.38)$$

$\gamma_E = 0.577\ldots$ the Euler–Mascheroni number, one concludes that ϕ_{qt} has a positive curvature at $k = 0$. The large k behavior can be inferred from the expansion

$$\psi_{di}(1 + ik) + \psi_{di}(1 - ik) = \log(1 + k^2) - \frac{1}{1 + k^2} + \mathcal{O}(k^{-4}). \qquad (14.39)$$

For the semi-classical limit, we set $\hbar k = w$ and then

$$\lim_{\hbar \to 0} \phi_{qt}(\hbar^{-1}w) = \log w^2 = \phi_{to}(w), \qquad (14.40)$$

as it should be.

In 2010, Kozlowski and Teschner obtained a remarkable result on the spectral properties of the chain, which comes handy now. They fix the eigenvalue E_1 and establish that for each vector $(\tau_1, \ldots, \tau_N)$ there is another vector $(\delta_1, \ldots, \delta_N)$ such that

$$\sum_{j=1}^{N} (\tau_j)^m = \sum_{j=1}^{N} (\hbar\delta_j)^m + \mathcal{O}(e^{-N}) \qquad (14.41)$$

with $m = 1, \ldots, N$. In addition, for every ordered set $\{n_1 < \cdots < n_N, n_j \in \mathbb{Z}, \}$, the roots $(\delta_1, \ldots, \delta_N)$ satisfy the identity

$$2\pi n_j = N\delta_j \log \hbar^2 + i \log \zeta - \sum_{i=1}^{N} i \log \frac{\Gamma(1 + i(\delta_j - \delta_i))}{\Gamma(1 - i(\delta_j - \delta_i))} + \mathcal{O}(e^{-N}). \quad (14.42)$$

This expression already has the flavor of (13.14) for the δ-Bose gas, except for the model dependent phase shift, of course. For the δ-Bose gas, (13.14) and (13.17) are a strict identities, no error term, whereas for the quantum Toda lattice, (14.41) and (14.42) hold approximately for large N.

Actually, for both error terms, there is an explicit identity. So far, the exponential bound $\mathcal{O}(e^{-N})$ has been established only for particular cases.

14.3 GGE and hydrodynamics

Given the input from (14.42) and assuming $n_1 < \cdots < n_N$, with small modifications one can follow the Yang–Yang scheme from the δ-Bose gas. By definition, the parameter ζ has unit length, $|\zeta| = 1$, and hence drops out as $N \to \infty$. The empirical density (14.33) satisfies the constraints

$$\varrho_N(w) \geqslant 0, \quad \int_{\mathbb{R}} dw \varrho_N(w) = 1, \quad \int_{\mathbb{R}} dw \varrho_N(w)w = N^{-1}E_1 = e_1. \quad (14.43)$$

Omitting both error terms, Eq. (14.42) becomes

$$\frac{1}{N}2\pi n_j = \frac{1}{N}\sum_{i=1}^{N}\theta_{\mathrm{qt}}(\delta_j - \delta_i) + e_1. \quad (14.44)$$

The GGE is defined through the weight

$$\exp\left[-\sum_{j=1}^{N}V(\delta_j)\right] \quad (14.45)$$

with confining potential V. Following the analysis of the δ-Bose gas, for quantum Toda, the free energy functional per particle reads

$$\mathcal{F}_{\mathrm{qt}}^{\circ}(\varrho) = \int_{\mathbb{R}} dw\big(\varrho V + \varrho\log\varrho + \varrho_{\mathrm{h}}\log\varrho_{\mathrm{h}} - (\varrho + \varrho_{\mathrm{h}})\log(\varrho + \varrho_{\mathrm{h}})\big). \quad (14.46)$$

This functional has to be minimized under two constraints. First, number and momentum,

$$\varrho \geqslant 0, \quad \int_{\mathbb{R}} dw \varrho(w) = 1, \quad \int_{\mathbb{R}} dw \varrho(w)w = e_1. \quad (14.47)$$

The second constraint comes from Eq. (14.44). Defining the operator T through the convolution with the quantum Toda scattering shift,

$$Tf(w) = \frac{1}{2\pi}\int_{\mathbb{R}} dw'\phi_{\mathrm{qt}}(w - w')f(w'), \quad (14.48)$$

the additional constraint can be written as

$$\varrho + \varrho_{\mathrm{h}} = T\varrho. \quad (14.49)$$

The constraint for the average momentum can be absorbed into a chemical potential μ_1, thereby adjusting the confining potential $V(w)$ to $V(w) - \mu_1 w$.

The variational problem appears to be identical to the one for the δ-Bose gas with volume $\ell = 0$, compare with (13.38). This should come as no surprise, since we used periodic boundary conditions with zero tilt. In this case, the stretch vanishes by construction. For the classical Toda lattice with this particular boundary conditions, the positions have the statistics of a random walk with step size of order 1 and zero bias, hence a fluctuating volume of order $\sqrt{N}$. For the classical chain, on top of the Flaschka invariants based on the Lax matrix, stretch is also conserved, respectively the particle density in the fluid picture. But so far we have not identified a corresponding control parameter in the quantum model. Still, physically, the stretch must be included in a proper hydrodynamic description. Clearly, one option is the cell Hamiltonian $H_{\text{cell},N}$, see Eq. (2.10). Since this Hamiltonian is written in canonical coordinates, the conventional quantization can be used which yields

$$H_{\text{cell},N,\nu} = -\sum_{j=1}^{N} \tfrac{1}{2}\hbar^2 \partial_{x_j}^2 + \sum_{j=1}^{N-1} X_j + \mathrm{e}^{-\nu N} X_N. \tag{14.50}$$

Before adopting this choice, we have to ensure that the tilt is properly imposed. For this purpose, we introduce the unitary transformation

$$U\psi(x_1, \dots, x_N) = \psi(x_1 - \nu, \dots, x_N - N\nu) \tag{14.51}$$

thereby obtaining

$$U^* p_j U = p_j, \quad U^* X_j U = \mathrm{e}^{-\nu} X_j, \quad U^* X_N U = \mathrm{e}^{\nu(N-1)} X_N, \tag{14.52}$$

and

$$U^* H_{\text{cell},N,\nu} U = H_{\text{qt},N,\nu} = -\sum_{j=1}^{N} \tfrac{1}{2}\hbar^2 \partial_{x_j}^2 + \sum_{j=1}^{N} \mathrm{e}^{-\nu} X_j. \tag{14.53}$$

Note that $H_{\text{qt},N,0} = H_{\text{qt},N}$. The boundary coupling can be spread homogeneously over the entire lattice. Due to invariance under spatial translations, the thermal operator $\exp(-H_{\text{cell},N,\nu})$ is not of trace class. However, for observables depending only on the internal coordinates, thermal average is well defined and denoted by $\langle \cdot \rangle_{N,\nu}$. Using unitary equivalence, one concludes

$$\langle (x_{j+1} - x_j) \rangle_{N,\nu} = \nu \tag{14.54}$$

for $j = 1, \dots, N-1$ and

$$\langle (x_1 - x_N) \rangle_{N,\nu} = -(N-1)\nu. \tag{14.55}$$

At the expense of a single huge jump, one indeed has induced a non-zero, constant stretch.

While convincing, one has to monitor how the higher order invariants are affected by such a modification. We set

$$M_\nu = \begin{pmatrix} 1 & 0 \\ 0 & \mathrm{e}^{-\nu N} \end{pmatrix}. \tag{14.56}$$

Then, the modified monodromy matrix reads

$$T_{N,\nu} = M_\nu T_N(u) \tag{14.57}$$

and

$$\hat{t}_{N,\nu}(u) = F_N^{\diamond}(u) + \mathrm{e}^{-\nu N} F_N^{\diamond\diamond}(u). \qquad (14.58)$$

Thus, the Hénon invariants of the modified Hamiltonian are still given by (14.18). Only, X_N is substituted everywhere by $\mathrm{e}^{-\nu N} X_N$. The same rule transcribes to the Flaschka invariants. Denoting the Flaschka invariants of $H_{\mathrm{cell},N,\nu}$ by $\hat{Q}_{\mathrm{cell},\nu}^{[m],N}$ and those of $H_{\mathrm{qt},N,\nu}$ by $\hat{Q}_{\mathrm{qt},\nu}^{[m],N}$, one arrives at

$$U^* \hat{Q}_{\mathrm{cell},\nu}^{[m],N} U = \hat{Q}_{\mathrm{qt},\nu}^{[m],N}. \qquad (14.59)$$

This equation teaches us that also under a generalized Gibbs ensemble for $H_{\mathrm{cell},N,\nu}$, the average stretch equals ν, except for the jump of $-\nu(N-1)$ at the bond $(N,1)$. Second, the generalized free energy can be computed from the Flaschka invariants of the lattice with coupling strength $\mathrm{e}^{-\nu}$. By scaling this means that in (14.36) the term $\log \hbar^2$ is replaced by $\log(\hbar^2 \mathrm{e}^{\nu})$. Thus, combining with the arguments leading to (14.44), the stretch modifies this relation to

$$2\pi n_j = \nu N \delta_j + \sum_{i=1}^{N} \theta_{\mathrm{qt}}(\delta_j - \delta_i) + N e_1. \qquad (14.60)$$

Up to the additional term coming from the fixed total momentum, we have obtained perfect analogy to the δ-Bose gas, compare with Eq. (13.14). However, in contrast to the δ-Bose gas, the stretch parameter νN can be negative, just as for the classical chain.

In our argument for the average currents of the δ-Bose gas, we never used the specific form of the scattering shift. For the quantum Toda lattice, the density is conserved and its current is the momentum, which itself is conserved. No extra work is required. To conclude, in the fluid picture, the hydrodynamic equations of the quantum Toda lattice read

$$\partial_t \rho_{\mathsf{p}}(x,t;v) + \partial_x \big(v^{\mathrm{eff}}(x,t;v)\rho_{\mathsf{p}}(x,t;v) \big) = 0. \qquad (14.61)$$

The effective velocity is again determined by the ratio $\varsigma_1^{\mathrm{dr}}/\varsigma_0^{\mathrm{dr}}$ as in (13.69). Of course, the appropriate dressing operator is the one for the quantum Toda lattice.

Notes and references

Section 14.0

Sutherland (1978) first discussed the integrability of the quantum Toda lattice, using a low-density approximation of the integrable quantum fluid with the $1/\sinh^2$ interaction potential. He developed the method of asymptotic Bethe ansatz, which uses properties of the eigenfunctions when particles are far apart, see Sutherland (1995). Finite volume has been introduced *ad hoc* by imposing periodic boundary conditions on the wave numbers. Sutherland studied TBA for the ground state. In a series of papers, Mertens and his group, Theodorakopoulos and Mertens (1983), Mertens (1984), Opper (1985), Hader and Mertens (1986), Gruner-Bauer and Mertens (1988), see also Theodorakopoulos (1984), Takayama and Ishikawa (1986), investigated TBA at non-zero temperatures and also the semi-classical classical limit. At the time, only thermal equilibrium was in focus. A most useful overview is provided by Siddharthan and Shastry (1997), see also Shastry and Young (2010). The review by Olshanetsky and Perelomov (1983) is an early summary on the more mathematical activities in the area of quantum integrable systems. Highly recommended is the instructive book on Beautiful Models by Sutherland (2004).

Gutzwiller (1980, 1981) studied the spectrum for few, $N = 3, 4$, particles, later being extended to an arbitrary number, see Gaudin and Pasquier (1992). Sklyanin (1985) achieved a crucial advance by linking the Toda lattice to the flourishing field of quantum integrable systems, with notions as Yang–Baxter equations and monodromy matrix. He developed the method of separation of variables, introducing the Baxter equation in this context. He also discussed the spectrum of eigenvalues in the limit of large N thereby supporting the more heuristic results in Sutherland (1978). Following a proposal by Nekrasov and Shatashvili (2010), Kozlowski and Teschner (2010) substantially advanced such methods.

Section 14.1

Our discussion is based on the contributions of Sklyanin (1985) and Siddharthan and Shastry (1997), see also Lüscher (1976). Identities as (14.9) are proved in Reed and Simon (1978), Section 17.

Section 14.2

A fairly explicit formula for the eigenfunctions of the Toda lattice has been obtained by Kharchev and Lebedev (1999, 2001). Their completeness is proved by An (2009). The results (14.41) and (14.42) are quoted from Kozlowski and Teschner (2010). The reader is invited to look up the precise statements. Further advances are reported by Kozlowski (2015). A more detailed account is his habilitation thesis Kozlowski (2015a).

Section 14.3

The kernel of the T operator of the δ-Bose gas decays to 0 for $|w| \to \infty$, while for the Toda lattice, the kernel diverges logarithmically. Hence, one has to check whether the techniques from Dorlas *et al.* (1993) still carry through. Sklyanin (1985) introduced already the parameter which we call ν and established that ν appears in the ground state TBA as the chemical potential dual to the density. The analysis of Kozlowski and Teschner (2010) starts with the Hamiltonian $H_{\mathrm{qt},N,\nu}$ from the outset. The parameter ν then naturally appears at the correct location. The short-cut discussed in the main text seems to be new.

Comparing (14.60) with (13.14), complete correspondence is noted. However, θ_{li} is strictly increasing, whereas θ_{qt} has a decreasing and increasing branch. Thus, uniqueness of solutions might be a further difficulty for an asymptotic analysis.

Chapter 15

Beyond the Euler Time Scale

As a common experience, sound waves in air propagate ballistically. When air is confined to a thin tube, an initial perturbation maintains its shape and travels with the speed of sound. There could be dispersive effects due to a nonlinear energy–momentum relation. The respective dynamics would be still time-reversible, no entropy is produced. But on longer time scales irreversible sound damping sets in, a phenomenon which can be traced back to molecular disorder. Very commonly, damping is modeled by a multiplicative diffusive factor as $\exp[-Dk^2|t|]$ in wave number space, D being the diffusion coefficient. On the level of hydrodynamic equations, dissipation amounts to the Navier–Stokes correction of Euler equations.

It is not at all obvious whether such a conventional picture extends to integrable systems. Again, the hard rod fluid serves as a convenient guiding example. Restricting to ballistic spacetime scales, the first step is to consider the motion of a tracer quasiparticle when the fluid is initially in some GGE. To lowest order, the tracer acquires an effective velocity, v^{eff}, through collisions with fluid quasiparticles. The effective velocity arises from summing over many approximately independent collisions, thereby leading to a law of large numbers. As well known from the theory of independent random variables, a more precise description would be a central limit theorem of the form

$$q_{\mathrm{tr}}(t) \simeq q_{\mathrm{tr}}(0) + v^{\mathrm{eff}}t + (D_{\mathrm{tr}})^{1/2}b(t), \tag{15.1}$$

where $q_{\mathrm{tr}}(t)$ is the position of the tracer quasiparticle, $b(t)$ a standard Brownian motion, and D_{tr} the diffusion constant, which depends on the particular GGE. For a generic integrable system, approximate statistical independence is less granted. But, as long as the notion of a tracer quasiparticle is defined, the asymptotics (15.1) is expected to be valid.

Connecting such a motion to the conserved fields leads to the Navier–Stokes correction, second order in ∂_x, whose analytical form still has to be worked out. To achieve such a goal, we follow a standard strategy. The initial attempt is to find out diffusive corrections for the propagation of a small perturbation away from the spatially homogeneous GGE. This will be soft mathematics, merely providing a general framework and even not yet distinguishing between integrable and nonintegrable systems. The result is then written in a form, for which the step from global GGE to local GGE will be compelling.

15.1 General framework

Despite the header, we explain the scheme using the Toda lattice as an example. The core of the method consists of suitable sum rules that can be written for other integrable models as well. As discussed in Section 7.1, the basic spacetime field–field correlator is

$$S_{m,n}(j,t) = \langle Q_j^{[m]}(t) Q_0^{[n]}(0) \rangle_{\mathrm{gg}}^{\mathrm{c}}. \tag{15.2}$$

Here, $\langle \cdot \rangle_{\mathrm{gg}}$ stands for the expectation in some GGE, which remains fixed throughout our discussion. By Lieb–Robinson-type bounds, for fixed t, the correlator $S_{m,n}(j,t)$ decays exponentially outside the sound cone. So j-summations are under control, while the time integration will turn out to be more delicate. To exploit spacetime symmetry, we redefine

$$Q_j^{[0]} = \tfrac{1}{2}(r_{j-1} + r_j). \tag{15.3}$$

To avoid too many extra symbols, for this section only, the basic correlator is still denoted by S, taking (15.3) into account.

⧫ *Spacetime inversion*: By spacetime stationarity, one concludes

$$S_{m,n}(j,t) = S_{n,m}(-j,-t). \tag{15.4}$$

In fact, the stronger property

$$S_{m,n}(j,t) = S_{m,n}(-j,-t) \tag{15.5}$$

holds by using invariance under spacetime inversion.
We define time reversal through

$$\mathcal{R}_{\mathrm{tr}} : (r,p) \mapsto (r,-p) \tag{15.6}$$

and space inversion by

$$\mathcal{R}_{\mathrm{si}} : (r, p) \mapsto (\tilde{r}, \tilde{p}), \quad \tilde{r}_j = r_{-j-1}, \; \tilde{p}_j = -p_{-j}. \tag{15.7}$$

Since r_j should be viewed as a bond variable, the inversion is relative to the origin. The equations of motion (2.2) generate a flow on phase space denoted by T_t, $T_t : (r, p) \mapsto (r(t), p(t))$. Using the flow equations, one concludes that

$$\mathcal{R}_{\mathrm{tr}} \circ T_t = T_{-t} \circ \mathcal{R}_{\mathrm{tr}} \tag{15.8}$$

and

$$\mathcal{R}_{\mathrm{si}} \circ T_t = T_t \circ \mathcal{R}_{\mathrm{si}}. \tag{15.9}$$

Spacetime inversion is then defined through

$$\mathcal{R}_{\mathrm{si}} \circ \mathcal{R}_{\mathrm{tr}} = \mathcal{R} : (r, p) \mapsto (\tilde{r}, \tilde{p}), \quad \tilde{r}_j = r_{-j-1}, \; \tilde{p}_j = p_{-j} \tag{15.10}$$

and hence

$$\mathcal{R} \circ T_t = T_{-t} \circ \mathcal{R}. \tag{15.11}$$

In the random walk summation for $Q_j^{[n]}$, see (2.18), to each path from j to j in n steps, there is a path reflected at level j. Hence,

$$Q_j^{[n]} \circ \mathcal{R} = Q_{-j}^{[n]}. \tag{15.12}$$

In particular, the GGE is invariant under $\mathcal{R}$. Combining (15.11) and (15.12), the claim (15.5) follows.

For the currents, spacetime inversion is slightly more complicated. We introduce the down and up currents

$$J_j^{[n]} = J_j^{[n]\downarrow} = \left(L^n L^{\downarrow} \right)_{j,j}, \quad J_j^{[n]\uparrow} = \left(L^n L^{\uparrow} \right)_{j,j}, \tag{15.13}$$

with $L^{\uparrow} = (L^{\downarrow})^{\mathrm{T}}$, see above Eq. (2.25). Then,

$$J_j^{[n]\downarrow} \circ \mathcal{R} = J_{-j}^{[n]\uparrow}. \tag{15.14}$$

Correspondingly, there are two current–current correlators

$$\Gamma_{m,n}^{\downarrow(\uparrow)}(j, t) = \langle J_j^{[m]\downarrow(\uparrow)}(t) J_0^{[n]\downarrow(\uparrow)}(0) \rangle_{\mathrm{gg}}^{\mathrm{c}} \tag{15.15}$$

and hence spacetime inversion implies

$$\Gamma_{m,n}^{\downarrow}(j, t) = \Gamma_{m,n}^{\uparrow}(-j, -t). \tag{15.16}$$

The down and up currents are presumably an *ad hoc* feature of the Toda lattice. ◆◆

Besides invariance under spacetime reversal, the correlator also satisfies the conservation law

$$\partial_t S_{m,n}(j,t) = -\partial_j B_{m,n}(j,t),$$ (15.17)

employing the difference operator $\partial_j f(j) = f(j+1) - f(j)$. $B(j,t)$ is the charge–current spacetime correlator

$$B_{m,n}(j,t) = \langle J_j^{[m]}(t) Q_0^{[n]}(0) \rangle_{gg}^c,$$ (15.18)

which has been encountered before, compare with (6.8). Using both properties, one derives identities, known as sum rules, for the lowest moments of $S(t)$.

For the *zeroth moment*, one obtains

$$\sum_{j \in \mathbb{Z}} S_{m,n}(j,t) = \sum_{j \in \mathbb{Z}} C_{m,n}(j) = C_{m,n},$$ (15.19)

where we used the conservation law (15.17) and the decay of $S_{m,n}(j,t)$ ensuring boundary terms to vanish. $C_{m,n}(j)$ is the static correlator,

$$C_{m,n}(j) = S_{m,n}(j,0),$$ (15.20)

and the spatially summed $C_{m,n}(j)$ is the matrix of static susceptibilities, $C_{m,n} = C_{n,m}$.

First moment: By (15.5), $C_{m,n}(j) = C_{m,n}(-j)$ and the sum rule

$$\sum_{j \in \mathbb{Z}} j C_{m,n}(j) = 0$$ (15.21)

follows. In addition, multiplying (15.17) with j and summing over j, one obtains

$$\partial_t \sum_{j \in \mathbb{Z}} j S_{m,n}(j,t) = \sum_{j \in \mathbb{Z}} B_{m,n}(j,t) = \sum_{j \in \mathbb{Z}} B_{m,n}(j,0) = B_{m,n} = (AC)_{m,n}$$ (15.22)

with the flux Jacobian A as encountered before, compare with (7.15). Hence,

$$\sum_{j \in \mathbb{Z}} j S_{m,n}(j,t) = (AC)_{m,n} t.$$ (15.23)

Using time stationarity of S, one reconfirms the symmetry of B, equivalently

$$AC = CA^{\mathrm{T}}.$$ (15.24)

Second moment: We pick some rapidly decaying test function f. Then,

$$\sum_{j\in\mathbb{Z}} f_j\big(Q_j^{[n]}(t) - Q_j^{[n]}(0)\big) = \int_0^t ds \sum_{j\in\mathbb{Z}} (\partial_j f_j) J_j^{[n]}(s). \tag{15.25}$$

Squaring and using translation invariance,

$$\sum_{j,j'\in\mathbb{Z}} f_j \tilde{f}_{j'} \big\langle \big(Q_j^{[m]}(t) - Q_j^{[m]}(0)\big)\big(Q_{j'}^{[n]}(t) - Q_{j'}^{[n]}(0)\big)\big\rangle_{gg}^c$$

$$= -\int_0^t ds \int_0^t ds' \sum_{j,j'\in\mathbb{Z}} (\partial_j^{\mathrm{T}}\partial_j f_j)\tilde{f}_{j'} \langle J_j^{[m]}(s) J_{j'}^{[n]}(s')\rangle_{gg}^c. \tag{15.26}$$

The central tool for controlling diffusive behavior is the spacetime current–current correlator defined by

$$\Gamma_{m,n}(j,t) = \langle J_j^{[m]}(t) J_0^{[n]}(0)\rangle_{gg}^c = \Gamma_{m,n}^{\downarrow}(j,t) \tag{15.27}$$

and the corresponding total current–current correlation

$$\sum_{j\in\mathbb{Z}} \Gamma_{m,n}(j,t) = \Gamma_{m,n}(t). \tag{15.28}$$

Note that by (15.16), $\Gamma_{m,n}(t) = \Gamma_{m,n}^{\downarrow}(t) = \Gamma_{m,n}^{\uparrow}(-t) = \Gamma_{m,n}(-t)$ and thus $\Gamma_{m,n}(\infty) = \Gamma_{m,n}(-\infty)$. Setting $f_j = \tfrac{1}{2}j^2$, $\tilde{f}_j = \delta_{0,j}$ and using the spacetime inversion (15.5), one concludes that

$$\sum_{j\in\mathbb{Z}} j^2 \big(S_{m,n}(j,t) - S_{m,n}(j,0)\big)$$

$$= \Gamma_{m,n}(\infty)t^2 + \int_0^t ds \int_0^t ds' \big(\Gamma_{m,n}(s-s') - \Gamma_{m,n}(\infty)\big). \tag{15.29}$$

$\Gamma_{m,n}(\infty)$ is the *Drude weight*. If $\Gamma_{m,n}(t) - \Gamma_{m,n}(\infty)$ is integrable, one can define the *Onsager matrix*

$$\mathfrak{L}_{m,n} = \int_{\mathbb{R}} dt \big(\Gamma_{m,n}(t) - \Gamma_{m,n}(\infty)\big). \tag{15.30}$$

$\mathfrak{L}$ is a symmetric matrix with $\mathfrak{L} \geqslant 0$ as covariance matrix. Hence, for large times,

$$\sum_{j\in\mathbb{Z}} j^2 S_{m,n}(j,t) \simeq \Gamma_{m,n}(\infty)t^2 + \mathfrak{L}_{m,n}|t|. \tag{15.31}$$

The reader may worry that we have lost our way, lots of definitions, and not even the diffusion matrix of interest. To appreciate our preparations, a useful example is the shear viscosity of a fluid. A standard scheme to experimentally define the shear viscosity, ν, is the Couette flow. One plate is fixed and the second plate, distance y and in parallel, is moved with constant velocity u. For small u, one finds the relation

$$\frac{F}{A} = \nu\frac{u}{y}. \tag{15.32}$$

Here, F/A is the pushing force per area acting on the moving plate. The shear viscosity reappears on a more sophisticated level in the Navier–Stokes equations. For example, solving these equations for the Couette flow confirms (15.32). But having the full equations, one can think of different set-ups to yield ν. In our context, one example would be the propagation of a small perturbation of an equilibrium state. Of course, experimental accessibility may vary. But, independently of the method, measured is always the same shear viscosity.

Returning to integrable systems, if such dissipative features are present at all, diffusion is expected to be governed by a high-dimensional matrix and *a priori* it is not so obvious whether and how familiar methods would apply. For the Euler time scale, we already identified the dynamic correlator as

$$\hat{S}_{m,n}(k,t) \simeq \left(e^{-\mathrm{i}kAt}C\right)_{m,n} \tag{15.33}$$

in wave number space, valid for small k and large t, kt fixed, compare with (7.21) and (7.25). Commonly, diffusive corrections are added through

$$\hat{S}_{m,n}(k,t) \simeq \left(e^{-\mathrm{i}kAt-k^2D|t|}C\right)_{m,n}. \tag{15.34}$$

By definition, D is the infinite-dimensional *diffusion matrix*. In general, D is not symmetric. However, to describe dissipation, D is required to have non-negative eigenvalues. In generic examples, D has a few zero eigenvalues and hence there is no diffusive decay along some particular directions. But $[A, D] \neq 0$ and A mixes the modes such that $\hat{S}(k,t)$ still decays to 0 in the long time limit.

The first moment sum rule is obviously satisfied. Computing the second moment in the position space version of (15.34) and comparing with the sum rule (15.31), one obtains

$$A^2Ct^2 + DC|t| = \Gamma(\infty)t^2 + \mathfrak{L}|t|. \tag{15.35}$$

Hence, the Drude weight matrix is given by

$$\Gamma(\infty) = A^2C = ACA^{\mathrm{T}} = B\frac{1}{C}B. \tag{15.36}$$

The latter expression is a natural symmetric form, which emphasizes that the Drude weight is a property determined by static correlations only.

In addition, the diffusion matrix is obtained as

$$DC = CD^{\mathrm{T}} = \mathfrak{L}, \quad D = \mathfrak{L}C^{-1}, \tag{15.37}$$

which is the generalization of the *Onsager relation* to integrable systems. Since $\mathfrak{L} \geqslant 0$ and $C > 0$, D is a similarity transform of $\mathfrak{L}$ and thus has only non-negative eigenvalues.

** *Drude weight*: In the context of condensed matter physics, it is fairly common to equate a non-zero Drude weight with infinite conductivity, implicitly suggesting that no further properties have to be investigated. In fact, according to our discussion, a non-zero Drude weight simply indicates a ballistic component of the dynamic correlator. For nonintegrable systems in one dimension, the ballistic component consists of sharp δ-peaks moving with constant velocity on the Euler scale. Diffusion is then easily detected through the broadening of the peaks. For integrable systems, the ballistic component is extended and of the generic form $t^{-1}g(t^{-1}x)$ in position space with some smooth scaling function g. Now, diffusive corrections become harder to detect, unless special care is taken to suppress the broad ballistic background. An example is the domain wall discussed in Chapter 8. At the contact line, the profile jumps from $\rho_{\mathrm{n}-}$ to $\rho_{\mathrm{n}+}$. Diffusion will broaden the step to an error function. Another example is the XXZ chain at zero magnetization (half filling). At these specific parameters, the spin Drude weight vanishes and diffusive effects can be observed directly. **

15.2 Nonintegrable chains

Our preparations have been accomplished already in Section 7.1 and, in essence, we are left with duplicating from the previous subsection, recalling that now there are only three fields, hence $n = 0, 1, 2$. It will turn out to be instructive to dwell on further details. In Section 7.1, serif letters were used for 3×3 matrices so as to distinguish from the integrable case. The same convention is followed here. Also, in this section, to ease the comparison with the literature, we switch to the physical pressure, denoted as before by $\mathfrak{p}$, which means that P in Eq. (7.5) has to be substituted by $\mathfrak{p}\beta$. It is convenient to set $u = 0$. Then, the Gibbs state is invariant under time-reversal, which provides additional symmetries. The average stretch is $\nu = \langle r_0 \rangle_{\mathfrak{p},\beta}$ and average energy $\mathsf{e} = \langle e_0 \rangle_{\mathfrak{p},\beta}$, where we dropped the mean velocity as the parameter. By convexity, one can view $\mathfrak{p}$ as a function of

ν, e. The linearization matrix reads

$$
A = \begin{pmatrix} 0 & -1 & 0 \\ \partial_\nu \mathsf{p} & 0 & \partial_\mathsf{e}\mathsf{p} \\ 0 & \mathsf{p} & 0 \end{pmatrix}. \tag{15.38}
$$

The static correlator is given by

$$
C = \begin{pmatrix} \langle r_0 r_0 \rangle^\mathsf{c}_{\mathsf{p},\beta} & 0 & \langle r_0 V_{\mathrm{ch},0} \rangle^\mathsf{c}_{\mathsf{p},\beta} \\ 0 & \beta^{-1} & 0 \\ \langle r_0 V_{\mathrm{ch},0} \rangle^\mathsf{c}_{\mathsf{p},\beta} & 0 & \langle e_0 e_0 \rangle^\mathsf{c}_{\mathsf{p},\beta} \end{pmatrix} \tag{15.39}
$$

and the current–charge cross-correlation by

$$
B = \beta^{-1} \begin{pmatrix} 0 & -1 & 0 \\ -1 & 0 & \mathsf{p} \\ 0 & \mathsf{p} & 0 \end{pmatrix}. \tag{15.40}
$$

With this information, one obtains the Drude weight

$$
\Gamma(\infty) = \beta^{-1} \begin{pmatrix} 1 & 0 & -\mathsf{p} \\ 0 & c^2 & 0 \\ -\mathsf{p} & 0 & \mathsf{p}^2 \end{pmatrix}, \tag{15.41}
$$

where c is the isentropic speed of sound,

$$
c^2 = \frac{\beta \langle (e_0 + \mathsf{p}r_0)(e_0 + \mathsf{p}r_0) \rangle^\mathsf{c}_{\mathsf{p},\beta}}{\langle r_0 r_0 \rangle^\mathsf{c}_{\mathsf{p},\beta} \langle e_0 e_0 \rangle^\mathsf{c}_{\mathsf{p},\beta} - (\langle r_0 e_0 \rangle^\mathsf{c}_{\mathsf{p},\beta})^2}. \tag{15.42}
$$

Finally, we have to list the two dynamic characteristics introduced before. The Onsager matrix turns out as

$$
L = \begin{pmatrix} 0 & 0 & 0 \\ 0 & \sigma_\mathsf{p}^2 & 0 \\ 0 & 0 & \sigma_\mathsf{e}^2 \end{pmatrix}. \tag{15.43}
$$

The only nonvanishing matrix elements are momentum–momentum and energy–energy currents. The notation is supposed to indicate that they are related to the noise strength of a Ginzburg–Landau fluctuation theory. But the defining time integral (15.30) cannot be expected to yield an explicit expression. The 0s at the upper left borders of the matrix result from the stretch current being conserved. The 1, 2 matrix element vanishes since momentum and energy have opposite signs under time reversal. The resulting diffusion matrix is given by

$$
D = \begin{pmatrix} 0 & 0 & 0 \\ 0 & \sigma_{\mathsf{p}}^2 & 0 \\ \alpha\sigma_{\mathsf{e}}^2 & 0 & \sigma_{\mathsf{e}}^2 \end{pmatrix},
\tag{15.44}
$$

where $\alpha = -\langle r_0 V_{\mathrm{ch},0}\rangle_{\mathsf{p},\beta}^{\mathsf{c}}/\langle r_0 r_0\rangle_{\mathsf{p},\beta}^{\mathsf{c}}$. Note that L, and hence D, has a zero eigenvalue. Since $[A, D] \neq 0$, still $\hat{S}(k, t) = \exp[-\mathrm{i}kAt - k^2 D|t|]C \to 0$ as $t \to \infty$, except for $k = 0$. If one reinstalls a non-zero mean velocity u, the Gibbs state is no longer invariant under time reversal. By a Galilean transformation, one can still figure out the various matrices. As a particular consequence, while $L_{0,n} = 0$ for $n = 0, 1, 2$ because of stretch current conservation, the cross term no longer vanishes, $L_{1,2} \neq 0$. Also, the diffusion matrix picks up further non-zero entries.

In position space, on the Euler scale, $S(x, t)$ consists of three δ-peaks: the heat peak at rest and the two sound peaks traveling with velocity $\pm c$. The diffusion term broadens the peaks as $\sqrt{t}$. While this prediction looks innocent, it is completely off the track as noted already in the mid-1970s. We quietly assumed that $\Gamma_{1,1}(t)$ and $\Gamma_{2,2}(t)$ are integrable. A more refined theory arrives at the conclusion that both current correlations decay generically as $|t|^{-2/3}$, which is well confirmed by molecular dynamic simulations. Also, the shape functions differ from a Gaussian. As discussed at the end of Section 7.1, see Figure 7.1, the heat peak broadens as $t^{3/5}$ with a shape function whose Fourier transform is given by $\exp(-|k|^{5/3}|t|)$. In the theory of independent random variables, this distribution is known as stable symmetric Lévy law with exponent $\frac{5}{3}$. The two sound peaks broaden distinctly with the power law $t^{2/3}$. Computing their shape function is a more difficult enterprise and related to the Kardar–Parisi–Zhang nonlinear fluctuation theory. One considers the stochastic Burgers equation

$$
\partial_t u(x, t) = \partial_x \big(u(x, t)^2 + \partial_x u(x, t) + \xi(x, t) \big),
\tag{15.45}
$$

where $\xi(x, t)$ is spacetime white noise. Then, its stationary measures are Gaussian white noise of variance one and arbitrary mean. The scaling function for the sound peaks turn out to be identical to the stationary two-point function $\mathbb{E}\big(u(x, t)u(0, 0)\big)^{\mathsf{c}}$. Actually, the full picture is more complicated. Anharmonic chains are divided into three universality classes. For one class, both $\Gamma_{1,1}(t)$ and $\Gamma_{2,2}(t)$ have an integrable decay. In this case, the claims from the linear theory are fully confirmed.

15.3   Navier–Stokes equations

For the Toda lattice, there is currently no method to compute the Onsager matrix directly on the basis of the microscopic model. However, using form factor expansions, a concise formula for the Onsager matrix has become available for the δ-Bose gas. By itself, this is a surprising result, since to explicitly compute transport coefficients for many-particle systems is a rare exception. By the much emphasized analogy between integrable systems, one thereby arrives also at a firm prediction for the Toda fluid. We state here merely the result and discuss some of its consequences. The matrix $\mathfrak{L}_{m,n}$ uses the basis consisting of monomials ς_n. Structurally more transparent is a continuum basis, in which $\mathfrak{L}$ is represented by an integral kernel over $\mathbb{R}$. We first introduce the kernel

$$K(w_1, w_2) = \rho_\mathsf{p}(w_1)\rho_\mathsf{p}(w_2)|v^{\text{eff}}(w_1) - v^{\text{eff}}(w_2)||T^{\text{dr}}(w_1, w_2)|^2, \qquad (15.46)$$

the integral operator T^{dr} being defined through

$$T^{\text{dr}} = (1 - T\rho_\mathsf{n})^{-1}T, \qquad (15.47)$$

and its action on the constant function,

$$\kappa(w_1) = \int_{\mathbb{R}} \mathrm{d}w_2 K(w_1, w_2). \qquad (15.48)$$

Clearly, T^{dr} is a symmetric operator and thus also K. The claim is that, for general functions f, g on $\mathbb{R}$,

$$\langle f, \mathfrak{L}g \rangle = \frac{1}{2} \int_{\mathbb{R}^2} \mathrm{d}w_1 \mathrm{d}w_2 K(w_1, w_2) \left(\frac{f^{\text{dr}}(w_2)}{\rho_\mathsf{s}(w_2)} - \frac{f^{\text{dr}}(w_1)}{\rho_\mathsf{s}(w_1)} \right)$$
$$\times \left(\frac{g^{\text{dr}}(w_2)}{\rho_\mathsf{s}(w_2)} - \frac{g^{\text{dr}}(w_1)}{\rho_\mathsf{s}(w_1)} \right). \qquad (15.49)$$

As it should be, $\mathfrak{L}$ is symmetric and $\mathfrak{L} \geqslant 0$. To determine possible zero eigenvalues, we note that $K(w_1, w_2) > 0$. Hence, $\mathfrak{L}f = 0$ implies

$$f^{\text{dr}}(w) = c\rho_\mathsf{s}(w) \qquad (15.50)$$

with arbitrary $c \in \mathbb{R}$, which has $f = c\varsigma_0$ as only solution. The zero subspace of $\mathfrak{L}$ corresponds to the projector $|\varsigma_0\rangle\langle\varsigma_0|$.

The diffusion matrix is obtained from the Onsager matrix as

$$D = \mathfrak{L}C^{-1}. \qquad (15.51)$$

The C matrix for the Toda lattice is stated in (7.26), respectively (7.29). The corresponding matrices for the Toda fluid are obtained by setting $\nu = 1$

and dropping the comoving terms $q_n\varsigma_0$. The result reads

$$C = (1 - \rho_n T)^{-1}\rho_p(1 - T\rho_n)^{-1}. \tag{15.52}$$

Hence,

$$C^{-1} = (1 - T\rho_n)\frac{1}{\rho_p}(1 - \rho_n T) \tag{15.53}$$

and

$$(C^{-1}g)^{\mathrm{dr}} = \frac{1}{\rho_p}(1 - \rho_n T)g. \tag{15.54}$$

Note that the zero eigenvalue of D has the left eigenvector ς_0 and the right eigenvector $C\varsigma_0$.

For applications, it will be more convenient to also have the integral kernel of D available. In order to separate into a diagonal multiplication operator and smooth off-diagonal integral kernel, we set

$$Uf(w) = \rho_s(w)^{-1}((1 - T\rho_n)^{-1}f)(w) \tag{15.55}$$

and

$$UC^{-1}g(w) = (\rho_s(w)\rho_p(w))^{-1}((1 - \rho_n T)g)(w). \tag{15.56}$$

Using the symmetry of K, one arrives at

$$\langle f, Dg \rangle = \int_{\mathbb{R}^2} dw_1 dw_2 K(w_1, w_2)(Uf(w_1)UC^{-1}g(w_1)$$

$$-Uf(w_1)UC^{-1}g(w_2)). \tag{15.57}$$

Setting $\bar{K} = U^{\mathrm{T}}KUC^{-1}$, it follows that

$$D(w_1, w_2) = -\bar{K}(w_1, w_2) + \delta(w_1 - w_2)U^{\mathrm{T}}\kappa UC^{-1}(w_1, w_1), \tag{15.58}$$

where κ is regarded as the multiplication operator. Since the left eigenvector of D is ς_0, if $\bar{K}(w_1, w_2) \geqslant 0$ and also $U^{\mathrm{T}}\kappa UC^{-1}(w_1, w_1) \geqslant 0$, then the operator $-D$ has the structure of the generator of a time-continuous Markov jump process.

A control check of (15.49) is easily performed by working out the case of hard rods. First, there is an exact expression for the structure function, here denoted by $\hat{S}^{\mathrm{hr}}_{m,n}(k, t)$, and its asymptotic behavior (15.34) has been confirmed. Using a sum rule as in (15.29), the total current correlator $\Gamma_{m,n}(t)$ can be obtained, with the result of being proportional to $\delta(t)$.

Thereby, the time integral trivializes and the Onsager matrix for hard rods is obtained as

$$\mathfrak{L}_{\mathrm{hr}}(w_1, w_2) = (a\bar{\rho})^2 (1 - a\bar{\rho})^{-1} \big(\delta(w_1 - w_2) r(w_1) - |w_1 - w_2| h(w_1) h(w_2) \big),$$
(15.59)

where $r(w) = \int_{\mathbb{R}} dw' h(w') |w - w'|$. Hence, the diffusion matrix becomes

$$D_{\mathrm{hr}}(w_1, w_2) = a(a\bar{\rho})(1 - a\bar{\rho})^{-1} \big(\delta(w_1 - w_2) r(w_1) - h(w_1) |w_1 - w_2| \big).$$
(15.60)

Second, in (15.49) and (15.57), one inserts the various explicit kernels and functions for hard rods, compare with Chapter 5. The resulting expressions are then in agreement with (15.59) and (15.60).

Based on our input, in analogy to classical fluids, the Navier–Stokes-type equation of the Toda fluid is given by

$$\partial_t \rho_{\mathsf{p}}(x, t; v) + \partial_x \big(v^{\mathrm{eff}}(x, t; v) \rho_{\mathsf{p}}(x, t; v) \big) = \partial_x D \partial_x \rho_{\mathsf{p}}(x, t; v) \qquad (15.61)$$

with diffusion matrix D of (15.58). The transformation to quasilinear form can still be carried out and yields

$$\partial_t \rho_{\mathsf{n}}(x, t) + v^{\mathrm{eff}}(x, t) \partial_x \rho_{\mathsf{n}}(x, t) = \rho_{\mathsf{s}}(x, t)^{-1} (1 - \rho_{\mathsf{n}} T) \partial_x D \partial_x \rho_{\mathsf{p}}(x, t).$$
(15.62)

As a generic physical requirement, Eq. (15.62) should yield a positive entropy production. More precisely, the balance equation for the local entropy will have a flow term and a production term. By the second law of thermodynamics, the latter should be positive. The local entropy at (x, t) is defined by

$$s(x, t) = -\langle \rho_{\mathsf{p}}(x, t) \log \rho_{\mathsf{n}}(x, t) \rangle, \qquad (15.63)$$

where $\langle \cdot \rangle$ refers to the integral over the spectral parameter, compare with Eq. (9.56). For a while, we fix the spectral parameter w and differentiate as

$$-\partial_t (\rho_{\mathsf{s}} \rho_{\mathsf{n}} \log \rho_{\mathsf{n}}) = -(\log \rho_{\mathsf{n}}) \partial_t \rho_{\mathsf{p}} - \rho_{\mathsf{s}} \partial_t \rho_{\mathsf{n}} = (\log \rho_{\mathsf{n}}) \partial_x (v^{\mathrm{eff}} \rho_{\mathsf{p}}) + \rho_{\mathsf{s}} v^{\mathrm{eff}} \partial_x \rho_{\mathsf{n}}$$
$$- \big(\log \rho_{\mathsf{n}} + (1 - \rho_{\mathsf{n}} T) \big) \partial_x (D \partial_x \rho_{\mathsf{p}}), \qquad (15.64)$$

where (15.61) and (15.62) have been inserted. First-order derivative terms contribute to the flow term. The first and second summands on the right of (15.64) combine to

$$(\log \rho_{\mathsf{n}}) \partial_x (v^{\mathrm{eff}} \rho_{\mathsf{p}}) + \rho_{\mathsf{s}} v^{\mathrm{eff}} \partial_x \rho_{\mathsf{n}} = \partial_x \big((\log \rho_{\mathsf{n}}) v^{\mathrm{eff}} \rho_{\mathsf{p}} \big) - (\partial_x \log \rho_{\mathsf{n}}) v^{\mathrm{eff}} \rho_{\mathsf{p}}$$
$$+ \rho_{\mathsf{s}} v^{\mathrm{eff}} \partial_x \rho_{\mathsf{n}} = \partial_x \big((\log \rho_{\mathsf{n}}) v^{\mathrm{eff}} \rho_{\mathsf{p}} \big). \quad (15.65)$$

For the third summand, we move the leftmost ∂_x in front, as before. This yields a further contribution to the flow term as

$$-\partial_x\big((\log\rho_n + (1 - T\rho_n))D\partial_x\rho_p\big) \tag{15.66}$$

together with the dissipative term

$$\big(\rho_n^{-1}\partial_x\rho_n - T\partial_x\rho_n\big)\big(D\partial_x\rho_p\big). \tag{15.67}$$

Differentiating $\rho_n = \rho_p(1 + T\rho_p)^{-1}$, one arrives at the identity

$$\rho_n^{-1}\partial_x\rho_n = \rho_p^{-1}(1 - \rho_n T)\partial_x\rho_p \tag{15.68}$$

and notes that

$$(1 - T\rho_n)\rho_n^{-1}\partial_x\rho_n = (1 - T\rho_n)\rho_p^{-1}(1 - \rho_n T)\partial_x\rho_p = C^{-1}\partial_x\rho_p. \tag{15.69}$$

We now substitute $D = \mathfrak{L}C^{-1}$ and arrive at the dissipative term

$$C^{-1}(\partial_x\rho_p)\mathfrak{L}C^{-1}(\partial_x\rho_p). \tag{15.70}$$

Altogether, the entropy balance for the Toda fluid reads

$$\partial_t s(x,t) + \partial_x \mathsf{j}_s(x,t) = \sigma(x,t). \tag{15.71}$$

The entropy current is determined to

$$\mathsf{j}_s = -\langle(\log\rho_n)v^{\text{eff}}\rho_p\rangle + \langle\big(\log\rho_n + (1 - T\rho_n)\big)D\partial_x\rho_p\rangle. \tag{15.72}$$

For the entropy production, the very concise formula

$$\sigma = \langle(\partial_x\rho_p)C^{-1}\mathfrak{L}C^{-1}(\partial_x\rho_p)\rangle \tag{15.73}$$

is unveiled, both j_s and σ depending on spacetime (x,t).

Notes and references

Section 15.1

For general anharmonic chains, the sum rules are discussed in Mendl and Spohn (2015a). While time reversal is a standard item, I learned the use of spacetime reversal from De Nardis *et al.* (2018, 2019). For hard rods, an exact formula for the structure factor $\hat{S}_{m,n}(k,t)$ is derived by Lebowitz *et al.* (1968). With this input, the small k behavior is studied by Spohn (1982), in particular the bulk diffusion matrix and respective current correlations. The Navier–Stokes correction for hard rods is proved by Boldrighini and Suhov (1997), with the result in complete agreement with (15.60).

Pioneering work on the Drude weight is Castella *et al.* (1995) and Zotos (1999). For fluids, the necessity to subtract the Drude weight has been recognized in Green (1954), see Spohn (1991) for a textbook discussion. In condensed matter physics, mostly one refers to the Mazur (1969), in particular to the Mazur bound which in our context means to sum in (15.36) only over a restricted number of conserved fields. The general formula is stated in Doyon and Spohn (2017). Applying (15.36) naively to the XXZ chain, one would conclude that the spin Drude weight vanishes. In actual fact, the Drude weight vanishes only for $\Delta > 1$ in the standard units, while it is non-zero and nowhere continuous for $0 \leqslant \Delta \leqslant 1$. The resolution can be traced back on our insistence on strictly local conservation laws. In actual fact, the XXZ chain has additional conservation laws whose densities have exponential tails and thus contribute to hydrodynamics, in particular to the Drude weight. We refer to Mierzejewski *et al.* (2015), Ilievski and De Nardis (2017), and the recent review Ilievski (2022) for more details.

Section 15.2

Nonintegrable anharmonic chains are studied at length by Spohn (2014), where the connection to the three-component stochastic Burgers equation, alias KPZ equation, is also explained. The dynamical correlator is confirmed through molecular dynamics with hard shoulder and other interaction potentials, see Mendl and Spohn (2014). Also, the nonintegrable decay of the total current–current correlation is convincingly observed in Mendl and Spohn (2015a). The hydrodynamic limit of anharmonic chains is studied by Bernardin and Olla (2020). Transport fluctuations are investigated in Myers *et al.* (2020). In Ganapa *et al.* (2021), molecular dynamics simulations are carried out, in which initially a huge amount of energy is deposited at the origin. The results compare well with hydrodynamic predictions.

Section 15.3

This section is entirely based on De Nardis *et al.* (2018, 2019), where the Navier–Stokes corrections are obtained through form factor expansions. The simple-looking formula for the local entropy production seems to be new.

Bibliography

Arutyunov, G. (2019). *Elements of Classical and Quantum Integrable Systems*, (Springer Nature Switzerland).

Abanov, A., Doyon, B., Dubai, J., Kamenev, A. and Spohn, H., eds. (2022). Hydrodynamics of low-dimensional quantum systems, *J. Phys. A*, special issue.

Ablowitz, M. J., Kaup, D. J., Newell, A. C. and Segur, H. (1974). The inverse scattering transform – Fourier analysis for nonlinear problems, *Stud. Appl. Math.* **53**, 249.

Ablowitz, M. J. and Ladik, J. F. (1975). Nonlinear differential-difference equations, *J. Math. Phys.* **16**, 598.

Ablowitz, M. J. and Ladik, J. F. (1976). Nonlinear differential-difference equations and Fourier analysis, *J. Math. Phys.* **17**, 1011.

Ablowitz, M. J. and Segur, H. (1981). *Solitons and Inverse Scattering Transform*, (SIAM, Philadelphia, PA).

Ablowitz, M. J., Prinari, B. and Trubatch, A. D. (2004). *Discrete and Continuous Nonlinear Schrödinger Systems* (Cambridge University Press).

Adler, M. (1977). Some finite dimensional integrable systems and their scattering behaviour, *Comm. Math. Phys.* **55**, 195.

Airault, H., McKean, H. P. and Moser, J. (1977). Rational and elliptic solutions of the Korteweg–de Vries equation and a related many-body problem, *Comm. Pure Appl. Math.* **30**, 95.

Akemann, G., Baik, J. and Di Francesco, P. (2011). *The Oxford Handbook on Random Matrix Theory* (Oxford University Press).

Alba, V. and Calabrese, P. (2017). Entanglement and thermodynamics after a quantum quench in integrable systems, *PNAS* **114**, 7947.

Aldous, D. (1985). *Exchangeability and Related Topics*, Lecture Notes in Mathematics, Vol. 117 (Springer, Berlin).

Allez, R., Bouchaud, J. P. and Guionnet, A. (2012). Invariant β-ensembles and the Gauss–Wigner crossover, *Phys. Rev. Lett.* **109**, 094102.

Allez, R., Bouchaud, J. P., Majumdar, S. N. and Vivo, P. (2012). Invariant β-Wishart ensembles, crossover densities and asymptotic corrections to the Marčenko–Pastur law, *J. Phys. A* **46**, 015001.

An, D. (2009). Complete set of eigenfunctions of the quantum Toda chain, *Lett. Math. Phys.* **87**, 209.

Anderson, G. W., Guionnet, A. and Zeitouni, O. (2010). *An Introduction to Random Matrices* (Cambridge University Press).

Bastianello, A., Bertini, B., Doyon, B. and Vasseur, R., eds. (2022). Emergent hydrodynamics in integrable many-body systems, special issue, *J. Stat. Mech.* **2022**.

Bastianello, A., Doyon, B., Watts, G. and Yoshimura, T. (2018). Generalized hydrodynamics of classical integrable field theory: The sinh-Gordon model, *SciPost Phys.* **4**, 045.

Bazhanov, V., Dorey, P., Kajiwara, K. and Takasaki, K., eds. (2018). Special issue on fifty years of the Toda lattice, *J. Phys. A* **51**.

Bernard, D. and Doyon, B. (2016). Conformal field theory out of equilibrium: A review, *J. Stat. Mech.* **2016**, 064005.

Bernardin, C. and Olla, S. (2010). Non-equilibrium macroscopic dynamics of chains of anharmonic oscillators, in preparation, pdf available on Researchgate.

Bertini, B., Collura, M., De Nardis, J. and Fagotti, M. (2016). Transport in out-of-equilibrium XXZ chains: Exact profiles of charges and currents, *Phys. Rev. Lett.* **117**, 207201.

Bertini, B., Heidrich-Meisner, F., Karrasch, C., Prosen, T., Steinigeweg, R. and Znidaric, M. (2021). Finite-temperature transport in one-dimensional quantum lattice models, *Rev. Mod. Phys.* **93**, 025003.

Bogomolny, E., Giraud, O. and Schmit, C. (2011). Integrable random matrix ensembles, *Nonlinearity* **24**, 3179.

Boldrighini, C., Dobrushin, R. L. and Sukhov, Yu. M. (1983). One-dimensional hard rod caricature of hydrodynamics, *J. Stat. Phys.* **31**, 577.

Boldrighini, C. and Suhov, Yu. M. (1997). One-dimensional hard rod caricature of hydrodynamics: Navier–Stokes correction for locally-equilibrium initial states, *Comm. Math. Phys.* **189**, 577.

Boltzmann, L. (1868). Studien über das Gleichgewicht der lebendigen Kraft zwischen bewegten materiellen Punkten, *Wiener Berichte* **58**, 517.

Boltzmann, L. (1868a). Lösung eines mechanischen Problems, *Wiener Berichte* **58**, 1035.

Bonnemain, T., Doyon, B. and El, G. A. (2022). Generalized hydrodynamics of the KdV soliton gas, *J. Phys. A* **55**, 374004.

Borsi, M., Pozsgay, B. and Pristyák, L. (2020). Current operators in Bethe ansatz and generalized hydrodynamics: An exact quantum/classical correspondence, *Phys. Rev. X* **10**, 011054.

Borsi, M., Pozsgay, B. and Pristyák, L. (2021). Current operators in integrable models: A review, *J. Stat. Mech.* **2021**, 094001.

Bouchoule, I., Doyon, B. and Dubail, J. (2020). The effect of atom losses on the distribution of rapidities in the one-dimensional Bose gas, *SciPost Phys.* **9**, 044.

Bouchoule, I. and Dubail, J. (2022). Generalized hydrodynamics in the 1D Bose gas: Theory and experiments, *J. Stat. Mech.* **2022**, 014003.

Bourgade, P., Erdös, L. and Yau, H. T. (2014). Edge universality of beta ensembles, *Comm. Math. Phys.* **332**, 261.

Bourgain, J. (1994). Periodic nonlinear Schrödinger equation and invariant measures, *Comm. Math. Phys.* **166**, 1.

Bressan, A. (2013). Hyperbolic conservation laws: An illustrated tutorial. In: *Modelling and Optimisation of Flows on Networks*, eds. Ambrosio, L. *et al.*, Lecture Notes in Mathematics, Vol. 2062, pp. 157–245 (Springer, Berlin, Heidelberg).

Brollo, A. and Spohn, H. (2023). Particle scattering for the Ablowitz-Ladik discretization of the nonlinear Schrödinger equation, in preparation.

Bulchandani, V. B. (2017). On classical integrability of the hydrodynamics of quantum integrable systems, *J. Phys. A* **50**, 453203.

Bulchandani, V. B., Cao, X. and Moore, J. (2019). Kinetic theory of quantum and classical Toda lattices, *J. Phys. A* **52**, 33LT01.

Bulchandani, V. B. and Karrasch, C. (2019). Subdiffusive front scaling in interacting integrable models, *Phys. Rev. B* **99**, 121410(R).

Bulchandani, V. B., Kulkarni, M., Moore, J. and Cao, X. (2021). Quasiparticle kinetic theory of Calogero particles, *J. Phys. A* **54**, 474001.

Bulchandani, V. B., Vasseur, R., Karrasch, C. and Moore, J. E. (2017). Solvable hydrodynamics of quantum integrable systems, *Phys. Rev. Lett.* **119**, 220604.

Calabrese, P., Essler, F. and Mussardo, G. (2016). Quantum integrability in out-of-equilibrium systems, special issue, *J. Stat. Mech.* **2016**, 064001–064011.

Calogero, F. (1971). Solution of the one-dimensional N-body problems with quadratic and/or inversely quadratic pair potentials, *J. Math. Phys.* **12**, 419.

Calogero, F.(1975). Exactly solvable one-dimensional many-body problems, *Lett. Nuovo Cimento* **13**, 411.

Calogero, F. (2001). *Classical Many-Body Problems Amenable to Exact Treatments* (Springer-Verlag, Heidelberg).

Cantero, M. J., Moral, L. and Velázquez, L. (2003). Five-diagonal matrices of orthogonal polynomials on the unit circle, *Linear Algebra Appl.* **362**, 29.

Cantero, M. J., Moral, L. and Velázquez, L. (2005). Minimal representations of unitary operators and orthogonal polynomials on the unit circle, *Linear Algebra Appl.* **408**, 40.

Cao, X., Bulchandani, V. B. and Moore, J. E. (2018). Incomplete thermalization from trap-induced integrability breaking: Lessons from classical hard rods, *Phys. Rev. Lett.* **120**, 164101.

Cao, X., Bulchandani, V. B. and Spohn, H. (2019). The GGE averaged currents of the classical Toda chain, *J. Phys. A* **52**, 495003.

Carbone, F., Dutykh, D. and El, G. A. (2016). Macroscopic dynamics of incoherent soliton ensembles: Soliton gas kinetics and direct numerical modelling, *Europhys. Lett.* **113**, 30003.

Castella, H., Zotos, X. and Prelovšek, P. (1995). Integrability and ideal conductance at finite temperatures, *Phys. Rev. Lett.* **74**, 972.

Castro-Alvaredo, O. A., Doyon, B. and Yoshimura, T. (2016). Emergent hydro-dynamics in integrable quantum systems out of equilibrium, *Phys. Rev. X* **6**, 041065.

Caux, J. S., Doyon, B., Dubail, J., Konik, R. and Yoshimura, T. (2019). Hydrodynamics of the interacting Bose gas in the quantum Newton cradle setup, *SciPost Phys.* **6**, 070.

Caux, J. S. and Mossel, J. (2011). Remarks on the notion of quantum integrability, *Journ. Stat. Mech.* **2011**, P02023.

Cépa, E. and Lépingle, D. (1997). Diffusing particles with electrostatic repulsion, *Probab. Theory Rel. Fields* **107**, 429.

Choquard, P. (2000). Classical and quantum partition functions of the Calogero–Moser–Sutherland model. In: *Calogero–Moser–Sutherland Models*, eds. J. F. van Diejen and L. Vinet, pp. 117–125 (Springer, New York).

Cohen, E. D. G., ed. (1975). *Fundamental Problems in Statistical Mechanics III*, Proceedings of the 1974 Wageningen Summer School (North-Holland, Amsterdam).

Coleman, S. R. and Mandula, J. (1967). All possible symmetries of the S-matrix, *Phys. Rev.* **159** 1251.

Collura, M., De Luca, A. and Viti, J. (2018). Analytic solution of the domain-wall nonequilibrium stationary state, *Phys. Rev. B* **97**, 081111(R).

Congy, T., El, G. A. and Roberti, G. (2021). Soliton gas in bidirectional dispersive hydrodynamics, *Phys. Rev. E* **103**, 042201.

Congy, T., El, G. A., Roberti, G. and Tovbis, A. (2023). Dispersive hydrodynamics of soliton condensates for the Korteweg–de Vries equation, *J. Nonlinear Science* **33**, 104.

Corwin, I. (2012). The Kardar–Parisi–Zhang equation and universality class, *Random Matrices Theory Appl.* **1**, 1130001.

Croydon, D. and Sasada, M. (2021). Generalized hydrodynamic limit for the box-ball system, *Comm. Math. Phys.* **383**, 427.

Cubero, A. C., Yoshimura, T. and Spohn, H. (2021). Form factors and generalized hydrodynamics for integrable systems, *J. Stat. Mech.* **2021**, 114002.

Cuccoli, A., Spicci, M., Tognetti, V. and Vaia, R. (1993). Dynamic correlations of the classical and quantum Toda lattices, *Phys. Rev. B* **47**, 7859.

Das, A., Damle, K., Dhar, A., Huse, D. A., Kulkarni, M., Mendl, C. B. and Spohn, H. (2020). Nonlinear fluctuating hydrodynamics for the classical XXZ spin chain, *J. Stat. Phys.* **180**, 238.

Das, A., Kulkarni, M., Spohn, H. and Dhar, A. (2019). Kardar–Parisi–Zhang scaling for an integrable lattice Landau–Lifshitz spin chain, *Phys. Rev. E* **100**, 042116.

Davies, B. (1990). Higher conservation laws for the quantum nonlinear Schrödinger equation, *Physica* **167**, 433.

Davies, B. and Korepin, V. E. (2011). Higher conservation laws for the quantum non-linear Schrödinger equation, arXiv:1109.6604.

Deift, P., Dubach, G. and Trogdon, T. (2019). *Universality in Numerical Computation*, Courant Lecture Notes, Wiley, forthcoming. Based on lectures at the Courant Institute.

De Luca, A. and Mussardo, G. (2016). Equilibration properties of classical integrable field theories, *J. Stat. Mech.* **2016**, 064011.

Dembo, A. and Zeitouni, O. (2010). *Large Deviations Techniques and Applications* (Springer-Verlag, Berlin).

De Nardis, J., Bernard, D. and Doyon, B. (2018). Hydrodynamic diffusion in integrable systems, *Phys. Rev. Lett.* **121**, 160603.

De Nardis, J., Bernard, D. and Doyon, B. (2019). Diffusion in generalized hydrodynamics and quasiparticle scattering, *SciPost Phys.* **6**, 049.

Deokule, A. (2022). Long time solution to the Euler equations of generalized hydrodynamics for periodic integrable systems, MSc Thesis, ICTS Bangalore.

Diederich, S. (1981). A conventional approach to dynamic correlations in the Toda lattice, *Phys. Lett. A* **85**, 233.

van Diejen, J. F. and Vinet, L., eds. (2012). *Calogero–Moser–Sutherland Models* (Springer, Heidelberg).

Dobrushin, R. L. (1974). Conditions for the absence of phase transitions in one-dimensional classical systems, *Math. USSR Sb.* **22**, 28.

Dobrushin, R. L. (1989). Caricatures of hydrodynamics. In: *IXth International Congress on Mathematical Physics*, eds. B. Simon, A. Truman, I.M. Davies, pp. 117–132 (Adam Hilger, Bristol).

Dorlas, T. C. (1993). Orthogonality and completeness of the Bethe ansatz eigenstates of the nonlinear Schrödinger model, *Comm. Math. Phys.* **154**, 347.

Dorlas, T. C., Lewis, J. T. and Pulé, J. V. (1989). The Yang–Yang thermodynamic formalism and large deviations, *Comm. Math. Phys.* **124**, 365.

Dotsenko, V. (2010). Replica Bethe ansatz derivation of the Tracy–Widom distribution of the free energy fluctuations in one-dimensional directed polymers, *J. Stat. Mech.* **2010**, P07010.

Doyon, B. (2008). Introduction to integrable quantum field theory, Lecture notes, Durham University.

Doyon, B. (2017). Thermalization and pseudolocality in extended quantum systems, *Comm. Math. Phys.* **351**, 155.

Doyon, B. (2018). Exact large-scale correlations in integrable systems out of equilibrium, *SciPost Phys.* **5**, 054.

Doyon, B. (2019). Generalised hydrodynamics of the classical Toda system, *J. Math. Phys.* **60**, 073302.

Doyon, B. (2020). Lecture notes on generalised hydrodynamics, *SciPost Phys.*, Lecture Notes 18.

Doyon, B., Dubail, J., Konik, R. and Yoshimura, T. (2017). Large-scale description of interacting one-dimensional Bose gases: Generalized hydrodynamics supersedes conventional hydrodynamics, *Phys. Rev. Lett.* **119**, 195301.

Doyon, B. and Myers, J. (2020). Fluctuations in ballistic transport from Euler hydrodynamics, *Ann. H. Poincaré* **21**, 255.

Doyon, B. and Spohn, H. (2017). Drude weight for the Lieb–Liniger Bose gas, *SciPost Phys.* **3**, 039.

Doyon, B. and Spohn, H. (2017a). Dynamics of hard rods with initial domain wall state, *J. Stat. Mech.* **2017**, 073210.

Doyon, B., Spohn, H. and Yoshimura, T. (2018). A geometric viewpoint on generalized hydrodynamics, *Nucl. Phys. B* **926**, 570.

Doyon, B., Yoshimura, T. and Caux, J. S. (2018). Soliton gases and generalized hydrodynamics, *Phys. Rev. Lett.* **120**, 045301.

Dumitriu, I. and Edelman, A. (2002). Matrix models for beta ensembles, *J. Math. Phys.* **43**, 5830.

Duy, T. K. (2018). On spectral measures of random Jacobi matrices, *Osaka J. Math.* **55**, 595.

Duy, T. K. and Shirai, T. (2015). The mean spectral measures of random Jacobi matrices related to Gaussian beta ensembles, *Electron. Commun. Probab.* **20**, 13.

Dyson, F. J. (1962). A Brownian-motion model for the eigenvalues of a random matrix, *J. Math. Phys.* **3**, 1191.

Dyson, F. J. (1969). Existence of a phase-transition in a one-dimensional Ising ferromagnet, *Comm. Math. Phys.* **12**, 91.

El, G. A. (2003). The thermodynamic limit of the Whitham equations, *Phys. Lett. A* **311**, 374.

El, G. A. (2021). Soliton gas in integrable dispersive hydrodynamics, *J. Stat. Mech.* **2021**, 114001.

El, G. A. and Kamchatnov, A. M. (2005). Kinetic equation for a dense soliton gas, *Phys. Rev. Lett.* **95**, 204101.

El, G. A., Kamchatnov, A. M., Pavlov, M. V. and Zykov, S. A. (2011). Kinetic equation for a soliton gas and its hydrodynamic reductions, *J. Nonlinear Sci.* **21**, 151.

Erdős, L., Yau, H. T. and Yin, J. (2012). Bulk universality for generalized Wigner matrices, *Probab. Theory Related Fields* **154**, 341.

Faddeev, L. D. (1995). Instructive history of the quantum inverse scattering method. In: *KdV95*, pp. 69–84 (Springer, Dordrecht).

Faddeev, L. D. (1996). How the algebraic Bethe ansatz works for integrable model, arXiv:hep-th/9605187.

Faddeev, L. D. and Takhtajan, L. (2007). *Hamiltonian Methods in the Theory of Solitons*, Classics in Mathematics (Springer, Heidelberg).

Fehér, L. (2013). Action-angle map and duality for the open Toda lattice in the perspective of Hamiltonian reduction, *Phys. Lett. A* **377**, 2917.

Ferguson, W. E., Flaschka, H. and McLaughlin, D. W. (1982). Normal modes for the Toda chain, *J. Comput. Phys.* **45**, 157.

Ferrari, P. A., Nguyen, C., Rolla, L. and Wang, M. (2021). Soliton decomposition of the box-ball system, *Forum Math. Sigma* **9**, E60.

Ferrari, P. A., Franceschini, C., Grevino, D. and Spohn, H. (2023). Hard rod hydrodynamics and the Levy Chentsov field, *ENSAIOS MATEMÁTICOS* **38**, 185.

Flaschka, H. (1974). The Toda lattice. I. Existence of integrals, *Phys. Rev. B* **9**, 1924.

Flaschka, H. (1974a). On the Toda lattice. II: Inverse-scattering solution, *Progr. Theor. Phys.* **51**, 703.

Flaschka, H. and McLaughlin, D. W. (1976). Canonically conjugate variables for the Korteweg–de Vries equation and the Toda lattice with periodic boundary conditions, *Progr. Theor. Phys.* **55**, 438.

Flaschka, H., Forest, M. G., and McLaughlin, D. W. (1980). Multiphase averaging and the inverse spectral solution of the Korteweg–de Vries equation, *Comm. Pure Appl. Math.* **33**, 739.

Ford, J., Stoddard, S. D. and Turner, J. S. (1973). On the integrability of the Toda lattice, *Progr. Theor. Phys.* **50**, 1547.

Forrester, P. J. (2010). *Log-Gases and Random Matrices*, LMS-34 (Princeton University Press).

Forrester, P. J. and Mazzuca, G. (2021). The classical β-ensembles with β proportional to $1/n$: From loop equations to Dyson's disordered chain, *J. Math. Phys.* **62**, 073505.

Forster, D. (1975). *Hydrodynamic Fluctuations, Broken Symmetry, and Correlation Functions* (Benjamin, Reading MA).

Franchini, F. (2017). *An Introduction to Integrable Techniques for One-Dimensional Quantum Systems*, Lecture Notes in Physics, Vol. 940 (Springer, Heidelberg).

Friedli, S. and Velenik, Y. (2017). *Statistical Mechanics of Lattice Systems: A Concrete Mathematical Introduction* (Cambridge University Press).

Fröhlich, J., Knowles, A., Schlein, B. and Sohinger, V. (2017). Gibbs measures of nonlinear Schrödinger equations as limits of many-body quantum states in dimensions $d \leq 3$, *Comm. Math. Phys.* **356**, 883.

Gamayun, O., Miao, Y. and Ilievski, E. (2019). Domain-wall dynamics in the Landau–Lifshitz magnet and the classical-quantum correspondence for spin transport, *Phys. Rev. B* **99**, 140301(R).

Ganapa, S., Chakraborti, S., Krapivsky, P. L. and Dhar, A. (2021). Blast in the one-dimensional cold gas: Comparison of microscopic simulations with hydrodynamic predictions, *J. Phys. Fluids* **33**, 087113

Gardner, C. S. (1971). Korteweg–de Vries equation and generalizations. IV. The Korteweg–de Vries equation as a Hamiltonian system, *J. Math. Phys.* **12**, 1548.

Gardner, C. S., Greene, J. M., Kruskal, M. D. and Miura, R. M. (1967). Method for solving the Korteweg–deVries equation, *Phys. Rev. Lett.* **19**, 1095.

Gaudin, M. (2014). *The Bethe Wavefunction*, translated by Caux J. S. (Cambridge University Press).

Gaudin, M. and Pasquier, V. (1992). The periodic Toda chain and a matrix generalization of the Bessel function recursion relations, *J. Phys. A* **25**, 5243.

Georgii, H. O. (1994). Large deviations and the equivalence of ensembles for Gibbsian particle systems with superstable interaction, *Prob. Theory Related Fields* **99**, 171.

Georgii, H. O. (1975). The equivalence of ensembles for classical systems of particles, *J. Stat. Phys.* **80**, 1341.

Gibbs, J. W. (1902). *Elementary Principles in Statistical Mechanics* (Charles Scribner's Sons, New York).

Girotti, M., Grava, T., Jenkins, R. and McLaughlin, K. (2021). Rigorous asymptotics of a KdV soliton gas, *Comm. Math. Phys.* **384**, 733.

Golinskiĭ, L. B. (2006). Schur flows and orthogonal polynomials on the unit circle, *Sbornik Math.* **197**, 1145.

Gon, A. K. and Kulkarni, M. (2019). Duality in a hyperbolic interaction model integrable even in a strong confinement: multi-soliton solutions and field theory, *J. Phys. A* **52**, 415201.

Grabowski, M. P. and Mathieu, P. (1994). Quantum integrals of motion for the Heisenberg spin chain, *Mod. Phys. Lett. A* **09**, 2197.

Grabowski, M. P. and Mathieu, P. (1995). Structure of the conservation laws in quantum integrable spin chains with short range interactions, *Ann. Phys.* **243**, 299.

Grava, T. and Mazzuca, G. (2023). Generalized Gibbs ensemble of the Ablowitz–Ladik lattice, Circular β-ensemble and double confluent Heun equation, *Comm. Math. Phys.* **399**, 1689.

Grava, T., Kriecherbauer, T., McLaughlin, K. and Mazzuca, G. (2021). Correlation functions for a chain of short range oscillators, *J. Stat. Phys.* **183**, 1.

Grébert, B. and Kappeler, T. (2014). *The Defocusing NLS Equation and Its Normal Form* (European Mathematical Society, Zürich).

Green, M. S. (1954). Markoff random processes and and the statistical mechanics of time-dependent phenomena. II. Irreversible processes in fluids, *J. Chem. Phys.* **22**, 398.

Grisi, R. and Schütz, G. (2011). Current symmetries for particle systems with several conservation laws, *J. Stat. Phys.* **145**, 1499.

Gruner-Bauer, P. and Mertens, F. G. (1988). Excitation spectrum of the Toda lattice for finite temperatures, *Z. Physik B Condens. Matter* **80**, 435.

Guionnet, A. and Memin, R. (2022). Large deviations for generalized Gibbs ensembles of the classical Toda chain, *EJP* **27**, 1.

Gutzwiller, M. C. (1980). The quantum mechanical Toda lattice, *Ann. Phys.* **124**, 347.

Gutzwiller, M. C. (1981). The quantum mechanical Toda lattice II, *Ann. Phys.* **133**, 304.

Hader, M. and Mertens, F. G. (1986). Thermodynamics of the Toda lattice, *J. Phys. A* **19**, 1913.

Hardy, A. and Lambert, G. (2021). CLT for circular beta-ensembles at high temperature, *J. Funct. Anal.* **280**, 108869.

Hénon, M. (1974). Integrals of the Toda lattice, *Phys. Rev. B* **9**, 1921.

Henrici, A. and Kappeler, T. (2008a). Global Birkhoff coordinates for the periodic Toda lattice, *Nonlinearity* **21**, 2713.

Henrici, A. and Kappeler, T. (2008b). Global action-angle variables for the periodic Toda lattice, *Int. Math. Res. Notes* **2008**, 1.

Henrici, A. and Kappeler, T. (2008c). Birkhoff normal form for the periodic Toda lattice, *Integr. Syst. Random Matrices Contemp. Math.* **458**, 11.

Hewitt, E. and Savage, J. L. (1985). Symmetric measures on Cartesian products, *Trans. Am. Math. Soc.* **80**, 470.

Hirota, R. (1973). Exact N-soliton solution of nonlinear lumped self-dual network equations, *J. Phys. Soc. Jpn.* **35**, 289.

Hohenberg, P. C. and Halperin, P. I. (1977). Theory of dynamical critical phenomena, *Rev. Mod. Phys.* **49**, 435.

Hubacher, A. (1989). Classical scattering theory in one dimension, *Comm. Math. Phys.* **123**, 353.

Iagolnitzer, D. (1978a). Factorization of the multiparticle S-matrix in two-dimensional space-time models, *Phys. Rev. D* **18**, 1275.

Iagolnitzer, D. (1978b). The multiparticle S matrix in two-dimensional space-time models, *Phys. Lett. B* **76**, 207.

Ilievski, E. (2022). Popcorn Drude weights from quantum symmetry, *J. Phys. A* **55**, 504005.

Ilievski, E. and De Nardis, J. (2017). Microscopic origin of ideal conductivity in integrable quantum models, *Phys. Rev. Lett.* **119**, 020602.

Ilievski, E. and De Nardis, J. (2017a). Ballistic transport in the one-dimensional Hubbard model: The hydrodynamic approach, *Phys. Rev. B* **96**, 081118.

Ilievski, E., De Nardis, J., Wouters, B., Caux, J. S., Essler, F. H. L. and Prosen, T. (2015). Complete generalized Gibbs ensembles in an interacting theory, *Phys. Rev. Lett.*, **115** 157201.

Ilievski, E., Medenjak, M., Prosen, T. and Zadnik, L. (2016). Quasilocal charges in integrable lattice systems *J. Stat. Mech.* **2016**, 064008.

Inozemtsev, V. I. (1983). New completely integrable multiparticle dynamic systems, *Joint Inst. Nucl. Res.*, JINR-R-4-83-664.

Israelsson, S. (2001). Asymptotic fluctuations of a particle system with singular interaction, *Stoch. Process. Appl.* **93**, 25.

Johansson, K. (1991). Condensation of a one-dimensional lattice gas, *Comm. Math. Phys.* **141**, 41.

Karchev, S. and Lebedev, D. (1999). Integral representation for the eigenfunctions of the quantum periodic Toda chain, *Lett. Math. Phys.* **50**, 53.

Karchev, S. and Lebedev, D. (2001). Integral representations for the eigenfunctions of the quantum open periodic Toda chains from the QISM formalism, *J. Phys. A* **34**, 2247.

Karevski, D. and Schütz, G. (2019). Charge–current correlation equalities for quantum systems far from equilibrium, *SciPost Phys.* **6**, 068.

Killip, R. and Nenciu, I. (2004). Matrix models for circular ensembles, *Int. Math. Res. Notes* **50**, 2665.

Killip, R. and Nenciu, I. (2007). CMV: The unitary analogue of Jacobi matrices, *Comm. Pure Appl. Math.* **60**, 1148.

Kinoshita, T., Wenger, T. and Weiss, D. S. (2006). A quantum Newton's cradle, *Nature* **440**, 900.

Korepin, V. E., Bogoliubov, N. M. and Izergin, A. G. (1997). *Quantum Inverse Scattering Method and Correlation Functions* (Cambridge University Press).

Koslov, I., Serban, D. and Vu, D. L. (2018). TBA and tree expansion. In: V. Dobrev, ed., *Quantum Theory and Symmetries with Lie Theory and Its Applications in Physics*, Vol. 2, pp. 77–99 (Springer Nature, Singapore).

Kozlowski, K. K. (2015). Unitarity of the SoV transform for the Toda chain, *Comm. Math. Phys.* **334**, 223.

Kozlowski, K. K. (2015a). Asymptotic analysis and quantum integrable models, Université de Bourgogne, arXiv:1508.06085.

Kozlowski, K. K. and Teschner, J. (2010). TBA for the Toda chain. In: *New Tends in Quantum Integrable Systems*, pp. 195–219 (World Scientific, Singapore).

Krajnik, Ž., Ilievski, E., Prosen, T. and Pasquier, V. (2021). Anisotropic Landau–Lifshitz model in discrete space-time, *SciPost Phys.* **11**, 051.

Kruskal, M. D., Miura, R. M., Gardner, C. S. and Zabusky, N. J. (1970). Korteweg–de Vries equation and generalizations. V. Uniqueness and nonexistence of polynomial conservation laws, *J. Math. Phys.* **11**, 952.

Krüger, H. and Teschl, G. (2009). Long-time asymptotics of the Toda lattice for decaying initial data revisited, *Rev. Math. Phys.* **21**, 61.

Kulkarni, M. and Polychronakos, A. (2017). Emergence of the Calogero family of models in external potentials: Duality, solitons and hydrodynamics, *J. Phys. A* **50**, 455202.

Kundu A. and Dhar, A. (2016). Equilibrium dynamical correlations in the Toda chain and other integrable models, *Phys. Rev. E* **94**, 062130.

Kuniba, A., Misguich, G. and Pasquier, V. (2020). Generalized hydrodynamics in box-ball system, *J. Phys. A* **53**, 404001.

Kuniba, A., Misguich G. and Pasquier, V. (2021). Generalized hydrodynamics in complete box-ball system for $U_q(\widehat{sl}_n)$, *SciPost Phys.* **10**, 095.

Landau, L. D. and Lifshitz, E. M. (1975). *Course in Theoretical Physics*, Vol. 6, Fluid Dynamics (Pergamon Press, Oxford).

Lanford, O. E. (1973). Entropy and equilibrium states in classical statistical mechanics. In: *Statistical Mechanics and Mathematical Problems*, Lenard, A., ed., pp. 1–113, Lecture Notes in Physics, Vol. 2 (Springer, Berlin).

Lanford, O. E., Lebowitz, J. L. and Lieb, E. H. (1977). Time evolution of infinite anharmonic systems, *J. Stat. Phys.* **16**, 453.

Lax, P. D. (1968). Integrals of nonlinear equations of evolution and solitary waves, *Comm. Pure Appl. Math.* **21**, 467.

Lebowitz, J. L., Percus, J. K. and Sykes, J. (1968). Time evolution of the total distribution function of a one-dimensional system of hard rods, *Phys. Rev.* **171**, 224.

Lebowitz, J. L., Rose, H. A. and Speer, E. R. (1988). Statistical mechanics of the nonlinear Schrödinger equation, *J. Stat. Phys.* **50**, 657.

Li, C., Zhou, T., Mazets, I., Stimming, H. P., Moeller, F. S., Zhu, Z., Zhai, Y., Xiong, W., Shen, X. and Schmiedmayer, J. (2020). Relaxation of bosons in one dimension and the onset of dimensional crossover, *SciPost Phys.* **9**, 058.

Lieb, E. H. (1963). Exact analysis of an interacting Bose gas. II. The excitation spectrum, *Phys. Rev.* **130**, 1616.

Lieb, E. H. and Liniger, W. (1963). Exact analysis of an interacting Bose gas. I. The general solution and the ground state, *Phys. Rev.* **130**, 1605.

Lüscher, M. (1976). Dynamical charges in the quantized massive Thirring model, *Nucl. Phys. B* **117**, 473.

Malvania, N., Zhang, Y., Le, Y., Dubail, J., Rigol, V. and Weiss, D. S. (2021). Generalized hydrodynamics in strongly interacting 1D Bose gases, *Science* **373**, 1129.

Manakov, S. V. (1974). Complete integrability and stochastization of discrete dynamical systems, *Z. Eksp. Ter. Fiz.* **67**, 543, translated in *Sov. Phys. JETP* **67**, 269.

Mazur, P. (1969). Non-ergodicity of phase functions in certain systems, *Physica* **43**, 533.

Mazzuca, G. and Memin, R. (2022). Large deviations for Ablowitz–Ladik lattice and the Schur flow, arXiv:2201.03429.

Mazzuca, G., Grava, T., Kriecherbauer, T., McLaughlin, K., Mendl, C. B. and Spohn, H. (2023). Equilibrium spacetime correlations of the Toda lattice on the hydrodynamic scale, arXiv:2301.02431.

Mendl, C. and Spohn, H. (2014). Equilibrium time-correlation functions for one-dimensional hard-point systems, *Phys. Rev. E* **90**, 012147.

Mendl, C. and Spohn, H. (2015). Low temperature dynamics of the one-dimensional discrete nonlinear Schrödinger equation, *J. Stat. Mech.* **2015**, P08028.

Mendl, C. and Spohn, H. (2015a). Current fluctuations for anharmonic chains in thermal equilibrium, *J. Stat. Mech.* **2015**, P03007.

Mendl, C. and Spohn, H. (2022). High-low pressure domain wall for the classical Toda lattice, *SciPost Phys.* **5**, 002.

Mertens, F. G. (1984). Bethe Ansatz for the Toda lattice: Ground state and excitations, *Z. Phys. B* **55**, 353.

Messer, J. and Spohn, H. (1982). Statistical mechanics of the isothermal Lane–Emden equation, *J. Stat. Phys.* **29**, 561.

Mestyán, M. and Alba, V. (2020). Molecular dynamics simulation of entanglement spreading in generalized hydrodynamics, *SciPost Phys.* **8**, 055.

Mierzejewski, M., Prelovšek, P. and Prosen, T. (2015). Identifying local and quasilocal conserved quantities in integrable systems, *Phys. Rev. Lett.* **114**, 140601.

Misguich, G., Mallick, K. and Krapivsky, P. L. (2017). Dynamics of the spin-Heisenberg chain initialized in a domain-wall state, *Phys. Rev. B* **96**, 195151.

Miura, R. M., Gardner, C. S. and Kruskal, M. D. (1968). Korteweg–de Vries equation and generalizations. II. Existence of conservation laws and constants of motion, *J. Math. Phys.* **9**, 1204.

Møller, F. S., Perfetto, G., Doyon, B. and Schmiedmayer, J. (2020). Euler-scale dynamical correlations in integrable systems with fluid motion, *SciPost Phys.* **3**, 016.

Møller, F. S. and Schmiedmayer, J. (2020). Introducing iFluid: A numerical framework for solving hydrodynamical equations in integrable models, *SciPost Phys.* **8**, 041.

Morrey, C. B. (1955). On the derivation of hydrodynamics from statistical mechanics, *Comm. Pure Appl. Math.* **8**, 279.

Moser, J. (1975). Finitely many mass points on the line under the influence of an exponential potential – An integrable system. In: *Dynamical Systems, Theory and Applications*, Lecture Note in Physics, Vol. 38, pp. 467–497 (Springer, Heidelberg).

Moser, J. (1975a). Three integrable Hamiltonian systems connected with isospetral deformations, *Adv. Math.* **16**, 197.

Mossel, J. and Caux, J. S. (2012). Generalized TBA and generalized Gibbs, *J. Phys. A* **45**, 255001.

Myers, J., Bhaseen, J., Harris, R. J. and Doyon, B. (2020). Transport fluctuations in integrable models out of equilibrium, *SciPost Phys.* **8**, 007.

Nakano, F. and Trinh, K. D. (2018). Gaussian beta ensembles at high temperature: Eigenvalue fluctuations and bulk statistics, *J. Stat. Phys.* **173**, 295.

Nenciu, I. (2005). Lax pairs for the Ablowitz–Ladik system via orthogonal polynomials on the unit circle, *Int. Math. Res. Notices* **2005**, 647.

Nenciu, I. (2006). CMV matrices in random matrix theory and integrable systems: A survey, *J. Phys. A* **39**, 8811.

Nozawa, Y. and Fukai, K. (2020). Explicit construction of local conserved quantities in the XYZ spin-$\frac{1}{2}$ chain, *Phys. Rev. Lett.* **125**, 090602.

Oh, T. and Quastel, J. (2013). On invariant Gibbs measures conditioned on mass and momentum, *J. Math. Soc. Jpn.* **65**, 13.

Olla, S., Varadhan, S. R. S. and Yau H. T. (1993). Hydrodynamical limit for a Hamiltonian system with weak noise, *Comm. Math. Phys.* **155**, 523.

Olshanetsky M. A. and Perelomov, A. M. (1981). Classical integrable finite-dimensional systems related to Lie algebras, *Phys. Rep.* **71**, 313.

Olshanetsky, M. A. and Perelomov, A. M. (1983). Quantum integrable systems related to Lie algebras, *Phys. Rep.* **94**, 313.

Opper, M. (1985). Analytical solution of the classical Bethe-ansatz solution for the Toda chain, *Phys. Lett. A* **112**, 201.

Orfanidis, S. J. (1978). Discrete sine-Gordon equations, *Phys. Rev. D* **18**, 3822.

Oxford, S. C. (1979). The Hamiltonian of the quantized non-linear Schrödinger equation, Ph.D. thesis, UC Los Angeles.

Parke, S. (1980). Absence of particle production and factorisation of the S-matrix in 1+1 dimensional models, *Nucl. Phys. B* **174**, 166.

Percus, J. K. (1969). Exact solutions of kinetics of a model classical fluid, *Phys. Fluids* **8**, 1560.

Philip, A. R. (2019). Soliton solutions to Calogero–Moser systems, Master thesis, Royal Institute of Technology, Stockholm.

Piroli, L. and Calabrese, P. (2015). Exact formulas for the form factors of local operators in the Lieb–Liniger model, *J. Phys. A* **48**, 454002.

Piroli, L., De Nardis, J., Collura, M., Bertini, B. and Fagotti, M. (2017). Transport in out-of-equilibrium XXZ chains: Nonballistic behavior and correlation functions, *Phys. Rev. B* **96**, 115124.

Polkovnikov, A., Sengupta, K., Silva, A. and Vengalattore, M. (2011). Colloquium: Nonequilibrium dynamics of closed interacting quantum system, *Rev. Mod. Phys.* **83**, 863.

Polychronakos, A. P. (1992). New integrable systems from unitary matrix models, *Phys. Lett. B* **277**, 102.

Polychronakos, A. P. (2006). The physics and mathematics of Calogero particles, *J. Phys. A* **39**, 12793.

Pozsgay, B. (2020). Algebraic construction of current operators in integrable spin chains, *Phys. Rev. Lett.* **125**, 070602.

Pozsgay, B. (2020a). Current operators in integrable spin chains: Lessons from long range deformations, *SciPost Phys.* **8**, 016.

Pristyák, L. and Pozsgay, B. (2023). Current mean values in the XYZ model, *SciPost Phys.* **14**, 158.

Prolhac, S. and Spohn, H. (2011). The propagator of the attractive delta-Bose gas in one dimension, *J. Math. Phys.* **52**, 122106.

Quastel, J. and Spohn, H. (2015). The one-dimensional KPZ equation and its universality class, *J. Stat. Phys.* **160**, 965.

Ramirez, J., Rider, B. and Virág, B. (2011). Beta ensembles, stochastic Airy spectrum, and a diffusion, *J. Am. Math. Soc.* **24**, 911.

Redor, I., Barthélemy, E., Michallet, H., Onorato, M. and Mordant, N. (2019). Experimental evidence of a hydrodynamic soliton gas, *Phys. Rev. Lett.* **122**, 214502.

Reed, M. and Simon, B. (1978). *Methods of Modern Mathematical Physics*, Vol. 4: Analysis of Operators (Academic Press, New York).

Rougerie, N. (2015). De Finetti theorems, mean-field limits and Bose–Einstein condensation, Lecture Notes, Collège de France and LMU München, arXiv: 1506.05263.

Ruelle, D. (1969). *Statistical Mechanics* (Benjamin, Reading MA).

Ruijsenaars, S. M. N. (1988). Action-angle maps and scattering theory for some finite-dimensional integrable systems: I. The pure soliton case, *Comm. Math. Phys.* **115**, 127.

Ruijsenaars, S. M. N. (1990). Relativistic Toda systems, *Comm. Math. Phys.* **133**, 217.

Ruijsenaars, S. M. N. (1995). Action-angle maps and scattering theory for some finite-dimensional integrable systems: Sutherland type systems and their duals, *Publ. RIMS Kyoto Univ.* **30**, 865.

Ruijsenaars, S. M. N. (1999). *Systems of Calogero–Moser Type*, CRM Series in Mathematical Physics "Particles and Fields" (Springer, New York).

Saff, E. B. and Totik, V. (1997). *Logarithmic Potentials with External Fields*, Vol. 316, Appendix B by Thomas Bloom, Grundlehren der Mathematischen Wissenschaften (Springer Verlag, Berlin).

Schemmer, M., Bouchoule, I., Doyon, B. and Dubail, J. (2019). Generalized hydrodynamics on an atom chip, *Phys. Rev. Lett.* **122**, 090601.

Schneider, T. (1983). Classical statistical mechanics of lattice dynamic model systems: Transfer integral and molecular-dynamics studies. In: *Statics and Dynamics of Nonlinear Systems*, G. Benedek, H. Bilz and R. Zeyher, eds., pp. 212–241 (Proceedings, Ettore Majorana Centre, Erice).

Schneider, T. and Stoll, E. (1980). Excitation spectrum of the Toda lattice: A molecular-dynamics study, *Phys. Rev. Lett.* **45**, 997.

Seoane, J. M. and Sanjuán, M. (2013). New developments in classical chaotic scattering, *Rep. Prog. Phys.* **76**, 016001.

Shastry, B. S. and Young, A. P. (2010). Dynamics of energy transport in a Toda ring, *Phys. Rev. B* **82**, 104306.

Shatashvili, S. L. and Nekrasov, N. A. (2010). Quantization of integrable systems and four dimensional gauge theories. In: *XVIth International Congress on Mathematical Physics*, pp. 265–289 (World Scientific, Singapore).

Shiraishi, N. (2019). Proof of the absence of local conserved quantities in the XYZ chain with a magnetic field, *EPL* **128**, 17002.

Siddharthan, R. and Shastry, B. S. (1997). Quantizing the Toda lattice, *Phys. Rev. B* **55**, 12196.

Simon, B. (2007). CMV matrices: Five years later, *J. Comput. Appl. Math.* **208**, 120.

Simon, B. (2016). *The Statistical Mechanics of Lattice Gases* (Princeton University Press).

Sklyanin, E. K. (1979). On complete integrability of the Landau–Lifshitz equation, LOMI Preprint No. E-3-1979 (unpublished).

Sklyanin, E. K. (1985). The quantum Toda chain. In: *Nonlinear Equations in Classical and Quantum Field Theory*, N. Sanchez, ed., pp. 196–233, Lecture Notes in Physics, Vol. 226 (Springer, Heidelberg).

Spohn, H. (1982). Hydrodynamical theory for equilibrium time correlation functions of hard rods, *Ann. Phys.* **141**, 353.

Spohn, H. (1987). Interacting Brownian particles: A study of Dyson's model. In: *Hydrodynamic Behavior of Interacting Particle Systems*, G. Papanicolaou, ed., IMA Volumes in Mathematics and its Applications 9 (Springer-Verlag, Berlin).

Spohn, H. (1991). *Large Scale Dynamics of Interacting Particles* (Springer-Verlag, Heidelberg).

Spohn, H. (2014). Nonlinear fluctuating hydrodynamics for anharmonic chains, *J. Stat. Phys.* **154**, 1191.

Spohn, H. (2017). The Kardar–Parisi–Zhang equation: A statistical physics perspective. In: *Les Houches Summer School July 2015 Session CIV Stochastic Processes and Random Matrices*, G. Schehr, A. Altland, Y.V. Fyodorov, N. O'Connell, and L.F. Cugliandolo, eds. (Oxford University Press).

Spohn, H. (2020). Generalized Gibbs ensembles of the classical Toda chain, *J. Stat. Phys.* **180**, 4.

Spohn, H. (2020). Ballistic space-time correlators of the classical Toda lattice, *J. Phys. A* **53**, 265004.

Spohn, H. (2020a). The collision rate ansatz for the classical Toda lattice, *Phys. Rev. E* **101**, 060103.

Spohn, H. (2022). Hydrodynamic equations for the Ablowitz–Ladik discretization of the nonlinear Schrödinger equation, *J. Math. Phys.* **63**, 033305.

Spohn, H. and Lebowitz, J. L. (1977). Stationary non-equilibrium states of infinite harmonic systems, *Comm. Math. Phys.* **54**, 97.

Sutherland, B. (1971). Quantum many-body problem in one dimension: Thermodynamics, *J. Math. Phys.* **12**, 251.

Sutherland, B. (1978). A brief history of the quantum soliton with new results on the quantization of the Toda lattice, *Rocky Mountain J. Math.* **8**, 413.

Sutherland, B. (1995). Confirmation of the asymptotic Bethe ansatz, *Phys. Rev. Lett.* **75**, 1248.

Sutherland, B. (2004). *Beautiful Models, 70 Years of Exactly Solved Quantum Many-Body Problems* (World Scientific, Singapore).

Takahashi, M. (1999). *Thermodynamics of One-Dimensional Solvable Models*, (Cambridge University Press, Cambridge).

Takahashi, D. and Satsuma, J. (1990). A soliton cellular automaton, *J. Phys. Soc. Jpn.* **59**, 3514.

Takayama, H. and Ishikawa, M. (1986). Classical thermodynamics of the Toda lattice as a classical limit of the two-component Bethe ansatz scheme, *Progr. Theor. Phys.* **76**, 820.

Takhtajan, L. A. (1977). Integration of the continuous Heisenberg spin chain through the inverse scattering method, *Phys. Lett. A* **64**, 235.

Tanaka, S. (1975). Korteweg–de Vries equation; asymptotic behavior of solutions, *Publ. Res. Inst. Math. Sci.* **10**, 367.

Tanaka, S. (1976). Periodic multi-soliton solutions of Korteweg–de Vries equation and Toda lattice, *Progr. Theor. Phys. Suppl.* **59**, 107.

Theodorakopoulos, N. (1984). Finite-temperature excitations of the classical Toda chain, *Phys. Rev. Lett.* **53**, 871.

Theodorakopoulos, N. and Mertens, F. G. (1983). Dynamics of the Toda lattice: A soliton-phonon phase-shift analysis, *Phys. Rev. B* **28**, 3512.

Toda, M. (1967a). Vibration of a chain with nonlinear interaction, *J. Phys. Soc. Jpn.* **22**, 431.

Toda, M. (1967b). Wave propagation in anharmonic lattices, *J. Phys. Soc. Jpn.* **23**, 501.

Toda, M. (1983). Nonlinear lattice and soliton theory, *IEEE Trans. Circ. Syst.* **30**, 542.

Toda, M. (1989). *Theory of Nonlinear Lattices*, 2nd edn. (Springer-Verlag, Heidelberg).

Trinh, H. D. and Trinh, K. D. (2021). Beta Jacobi ensembles and associated Jacobi polynomials, *J. Stat. Phys.* **185**, 1.

Van Diejen, J. F. (1995). Difference Calogero-Moser systems and finite Toda chains. *J. Math. Phys.* **36**, 1299.

Vasseur, R. and Moore, J. E. (2016). Nonequilibrium quantum dynamics and transport: From integrability to many-body localization, *J. Stat. Mech.* **2016**, 064010.

Venakides, S., Deift, P. and Oba, R. (1991). The Toda shock problem, *Comm. Pure Appl. Math.* **44**, 1171.

Vidmar, L. and Rigol, M. (2016). Generalized Gibbs ensemble in integrable lattice models, *J. Stat. Mech.* **2016**, 064007.

Vu, D. L. and Yoshimura, T. (2019). Equations of state in generalized hydrodynamics, *SciPost Phys.* **6**, 023.

Whitham, G. B. (1974). *Linear and Nonlinear Waves* (John Wiley & Sons, Inc.).

Yang, C. N. and Yang, C. P. (1969). Thermodynamics of a one-dimensional system of Bosons with repulsive delta-function interaction, *J. Math. Phys.* **10**, 1115.

Yoshimura, T. and Spohn, H. (2020). Collision rate ansatz for quantum integrable systems, *SciPost Phys.* **9**, 040.

Zakharov, V. E. (1971). Kinetic equation for solitons, *Sov. Phys. JETP* **33**, 538.

Zakharov, V. E. (2009). Turbulence in integrable systems, *Studies in Applied Mathematics* **122**, 219.

Zhidkov, P. (2001). On an infinite sequence of invariant measures for the cubic nonlinear Schrödinger equation, *Int. J. Math. and Math. Sci.* **28**, 375.

Zotos, X. (1999). Finite temperature Drude weight of the one-dimensional spin-Heisenberg model, *Phys. Rev. Lett.* **82**, 1764.

Zwerger, W. (2022). The Lieb–Liniger gas with cold atoms. In: *The Physics and Mathematics of Elliott Lieb*, R. L. Frank, A. Laptev, M. Lewin and R. Seiringer, eds., pp. 629–652 (EMS Press).

List of Symbols

conserved field $Q^{[n]}$, field density $Q_j^{[n]}$

current $J^{[n]}$, current density $J_j^{[n]}$

commutator $[\cdot,\cdot]$

correlator

S, S	spacetime charge-charge
C, C	static charge-charge
B, B	static charge-charge current
A, A	flux Jacobian

density ρ

ϱ	generic $\langle \varrho \rangle = 1$
ρ	generic $\langle \rho \rangle \neq 1$
ρ_n	number density, TBA solution
ρ_p	particle density, root density, TBA solution
ρ_s	space density, TBA solution
ρ_s	stationary solution, Dyson Brownian motion
ρ_Q	DOS Lax matrix
ρ_J	current DOS

effective velocity v^{eff}

free energy functionals $\mathcal{F}$

$\mathcal{F}^\circ(\varrho)$	$\langle \varrho \rangle = 1$
$\mathcal{F}(\rho)$	other constaints on $\langle \rho \rangle$
$\mathcal{F}^\bullet(\rho)$	no constraint

GGE average $\langle \cdot \rangle_{P,V}$, $\langle \cdot \rangle_{\rho_f,V}$

Hamiltonian H

H_{aa}	action-angle
H_{al}	Ablowitz-Ladik lattice
H_{ca}	Calogero fluid
H_{cell}	Toda cell
H_{ch}	generic chain
H_{cm}	Calogero-Moser model
H_{hr}	hard rod lattice
$H_{\mathrm{hr,fl}}$	hard rod fluid
H_{li}	Lieb-Liniger δ-Bose gas
H_{mec}	general pair interaction
H_{qt}	quantum Toda fluid
H_{to}	Toda lattice
$H_{\mathrm{to}}^{\diamond}$	open Toda lattice
$H_{\mathrm{to,f}}$	Toda fluid
H_{XYZ}	XYZ spin chain

Hilbert spaces $L^2(\mathbb{R}, \mathrm{d}x)$, $\ell_2(\mathbb{Z})$

integral over $\mathbb{R}$, resp. $[0, 2\pi]$, $\langle \cdot \rangle$

Lax matrix L_N, L

Lax pair (L_N, B_N), (L, B) (L, M)

partition function Z

Z_{de}	Dumitriu-Edelman
Z_{to}	Toda
Z_{li}	Lieb-Liniger
Z_{can}	canonical
Z_N	system size N

phase space Γ

$$\Gamma_N = \mathbb{R}^N \times \mathbb{R}^N$$
$$\Gamma_N^{\circ} = \mathbb{R}_+^N \times \mathbb{R}^N$$
$$\Gamma_N^{\diamond} = \mathbb{R}_+^{N-1} \times \mathbb{R}^N$$
$$\Gamma_N^{\triangleright} = \mathbb{W}_N \times \mathbb{R}^N$$

Poisson bracket $\{\cdot, \cdot\}$

potential
V	confining
V_{ca}	Calogero
V_{ch}	chain
V_{hr}	hard rod
V_{mec}	pair interaction

quasi-energy ε

scattering shift ϕ
ϕ_{ca}	Calogero fluid
ϕ_{hr}	hard rods
ϕ_{kv}	KdV solitons
ϕ_{qt}	quantum Toda
ϕ_{to}	Toda particles
ϕ_{tos}	Toda solitons

torus in N dimensions $\mathbb{T}^N$

Index